Synthesis of Flavonoids or Other Nature-Inspired Small Molecules

Contents

Synthesis of Flavonoids or Other Nature-Inspired Small Molecules

Editor

Giovanni Ribaudo

MDPI • Basel • Beijing • Wuhan • Barcelona • Belgrade • Manchester • Tokyo • Cluj • Tianjin

Editor
Giovanni Ribaudo
University of Brescia
Italy

Editorial Office
MDPI
St. Alban-Anlage 66
4052 Basel, Switzerland

This is a reprint of articles from the Special Issue published online in the open access journal *Molbank* (ISSN 1422-8599) (available at: https://www.mdpi.com/journal/molbank/special_issues/ Nat_Mol).

For citation purposes, cite each article independently as indicated on the article page online and as indicated below:

LastName, A.A.; LastName, B.B.; LastName, C.C. Article Title. *Journal Name* **Year**, *Volume Number*, Page Range.

ISBN 978-3-0365-3244-8 (Hbk)
ISBN 978-3-0365-3245-5 (PDF)

Cover image courtesy of Giovanni Ribaudo

About the Editor

Giovanni Ribaudo, PhD. My research activity is based on the combination of tools from synthetic, analytical (HPLC, NMR, mass spectrometry), and computational medicinal chemistry. My main research topics consist of the design and screening of small molecules interacting with peculiar DNA arrangements and in the study of nature-inspired phosphodiesterase (PDE) inhibitors targeting the central nervous system. I received my Master's degree in Medicinal Chemistry and Technology at the University of Padova (Italy) in 2011. In 2015, I graduated with my PhD in Pharmaceutical Sciences at the same Institution, after carrying out part of the research activity at the State University of New York in Albany (NY, USA). Between 2015 and 2019, I worked as a post-doc at the University of Padova in collaboration with a company operating in the field of the chemistry of natural compounds. In 2019, I joined the Department of Molecular and Translational Medicine of the University of Brescia.

MDPI

Editorial

Synthesis of Flavonoids or Other Nature-Inspired Small Molecules

Giovanni Ribaudo

Department of Molecular and Translational Medicine, University of Brescia, 25123 Brescia, Italy;
giovanni.ribaudo@unibs.it

Citation: Ribaudo, G. Synthesis of Flavonoids or Other Nature-Inspired Small Molecules. *Molbank* **2022**, *2022*, M1313. https://doi.org/10.3390/M1313

Received: 4 January 2022
Accepted: 7 January 2022
Published: 10 January 2022

Publisher's Note: MDPI stays neutral with regard to jurisdictional claims in published maps and institutional affiliations.

Natural compounds are endowed with an intriguing variety of scaffolds, functional groups and stereochemical properties. Natural evolution continuously gives rise to extremely complicated arrangements of polysubstituted rings and chiral centers. In addition to the interest that such molecules trigger from the point of view of synthetic organic chemistry per se, the world of natural compounds has been explored by medicinal chemists considering the valuable biological activities that they bear, inspiring the design and development of novel drugs.

The Special Issue of *Molbank*, entitled "Synthesis of Flavonoids or Other Nature-Inspired Small Molecules", was launched in Spring 2020, and collected 10 contributions by the end of 2021 with a wide geographical reach. The Guest Editor is grateful to all the authors that actively contributed to the Special Issue by submitting the results of their research activity in the field of the synthesis of natural and nature-inspired compounds.

Among small molecules of natural origin, flavonoids represent an outstanding chemical class. Together with their semi-synthetic derivatives, flavonoids have been extensively studied in light of their antioxidant, antiproliferative, antibacterial and anti-inflammatory activities, to name a few. Therefore, this Special Issue aimed to collect contributions related to the following topics:

- Chemistry of flavonoids, alkaloids and natural compounds;
- Semi-synthetic compounds;
- Heterocycles;
- Stereochemistry;
- Analytical chemistry and structure elucidation.

In fact, in addition to flavonoids, several other classes of natural or nature-inspired molecules, such as alkaloids and terpenoids, are particularly attractive from the synthetic and medicinal chemistry perspectives.

In the first paper, the preparation and characterization of a semi-synthetic hydrazone derivative of quercetin, studied as a phosphodiesterase inhibitor and synthesized through the single-step modification of the natural precursor, was reported by our research group from the University of Brescia (Italy) [1]. In the second paper, Prof. Unang Supratman and colleagues from Universitas Padjadjaran (Indonesia) described a polyoxygenated dimer-type xanthone isolated from the stem bark of *Garcinia porrecta* [2]. In the third paper, a new onoceranoid triterpene was isolated from the fruit peels of *Lansium domesticum* by Dr. Tri Mayanti and colleagues from Universitas Padjadjaran (Indonesia), and the compound was tested for its antiproliferative activity [3]. In the fourth paper, Dr. Mariia Nesterkina and colleagues from Odessa National Polytechnic University (Ukraine), described the synthesis of 2-propyl-N'-[1,7,7-trimethylbicyclo[2.2.1]hept-2-ylidene]pentanehydrazide, a camphor derivative with anticonvulsant properties [4]. In the fifth paper, the synthesis and characterization of a quinoline derivative were described by the group of Dr. Paolo Coghi and Prof. Vincent Kam Wai Wong from the Macau University of Science and Technology (China). Furthermore, the compound was tested for its antiproliferative activity against several cell lines [5]. In the sixth paper, the novel tetranortriterpenoid kokosanolide D was isolated from *Lansium domesticum* by Dr. Tri Mayanti and colleagues from Universitas

Padjadjaran (Indonesia), and the compound was characterized through a combination of spectroscopic techniques [6]. In the seventh paper, Dr. Mithun Rudrapal and colleagues from the Dhariwal Institute of Pharmaceutical Education & Research (India) described flavonoids from *Cordia dichotoma* and their antidiabetic activity, based on computational and experimental data [7]. In the eight paper, a novel curcumin analogue developed as a potential fluorescent dye for biological systems was synthesized and characterized by Prof. Marco Edilson Freire de Lima and colleagues from Universidade Federal Rural do Rio de Janeiro (Portugal) [8]. In the ninth paper, Dr. Diana Becerra, Dr. Juan-Carlos Castillo and colleagues from Universidad Pedagógica y Tecnológica de Colombia and Universidad de los Andes (Colombia) reported the synthesis of 2-oxo-2*H*-chromen-7-yl 4-chlorobenzoate obtained by *O*-acylation of 7-hydroxy-2*H*-chromen-2-one [9]. Eventually, in the tenth paper, lupeol extracted from *Bombax ceiba* was used by Dr. Thuc-Huy Duong from the Ho Chi Minh City University of Education (Vietnam), Dr. Jirapast Sichaem from Thammasat University (Thailand), and colleagues, as a starting material to produce semi-synthetic derivatives exhibiting α-glucosidase inhibitory activity [10].

All these interesting contributions demonstrate the flourishing interest of the international scientific community in the identification and optimization of novel synthetic routes for producing nature-inspired bioactive compounds. As a conclusive note, the Guest Editor would like to sincerely thank the Reviewers and the Assistant Editors for their valuable support and for having made the realization of this Special Issue possible.

Acknowledgments: The Guest Editor would like to sincerely thank Jessica Tecchio for her support.

Conflicts of Interest: The author declares no conflict of interest.

References

1. Gianoncelli, A.; Ongaro, A.; Zagotto, G.; Memo, M.; Ribaudo, G. 2-(3,4-Dihydroxyphenyl)-4-(2-(4-nitrophenyl)hydrazono)-4*H*-chromene-3,5,7-triol. *Molbank* **2020**, *2020*, M1144. [CrossRef]
2. Safitri, A.N.; Nurlelasari; Mayanti, T.; Darwati; Supratman, U. 5,5′-Oxybis(1,3,7-trihydroxy-9*H*-xanthen-9-one): A New Xanthone from the Stem Bark of *Garcinia porrecta* (Clusiaceae). *Molbank* **2020**, *2020*, M1153. [CrossRef]
3. Zulfikar; Putri, N.; Fajriah, S.; Yusuf, M.; Maharani, R.; Anshori, J.; Supratman, U.; Mayanti, T. 3-Hydroxy-8,14-secogammacera-7,14-dien-21-one: A New Onoceranoid Triterpenes from *Lansium domesticum* Corr. cv *Kokossan*. *Molbank* **2020**, *2020*, M1157. [CrossRef]
4. Nesterkina, M.; Barbalat, D.; Rakipov, I.; Kravchenko, I. 2-Propyl-*N′*-[1,7,7-trimethylbicyclo[2.2.1]hept-2-ylidene]pentanehydrazide. *Molbank* **2020**, *2020*, M1164. [CrossRef]
5. Coghi, P.; Ng, J.P.L.; Nasim, A.A.; Wong, V.K.W. *N*-[7-Chloro-4-[4-(phenoxymethyl)-1*H*-1,2,3-triazol-1-yl]quinoline]-acetamide. *Molbank* **2021**, *2021*, M1213. [CrossRef]
6. Fauzi, F.M.; Meilanie, S.R.; Zulfikar; Farabi, K.; Herlina, T.; Al Anshori, J.; Mayanti, T. Kokosanolide D: A New Tetranortriterpenoid from Fruit Peels of *Lansium domesticum* Corr. cv Kokossan. *Molbank* **2021**, *2021*, M1232. [CrossRef]
7. Hussain, N.; Kakoti, B.B.; Rudrapal, M.; Sarwa, K.K.; Celik, I.; Attah, E.I.; Khairnar, S.J.; Bhattacharya, S.; Sahoo, R.K.; Walode, S.G. Bioactive Antidiabetic Flavonoids from the Stem Bark of *Cordia dichotoma* Forst.: Identification, Docking and ADMET Studies. *Molbank* **2021**, *2021*, M1234. [CrossRef]
8. Chaves, O.A.; Sueth-Santiago, V.; Pinto, D.C.d.A.; Netto-Ferreira, J.C.; Decote-Ricardo, D.; de Lima, M.E.F. 2-Chloro-4,6-*Bis*{(*E*)-3-methoxy-4-[(4-methoxybenzyl)oxy]styryl}pyrimidine: Synthesis, Spectroscopic and Computational Evaluation. *Molbank* **2021**, *2021*, M1276. [CrossRef]
9. Becerra, D.; Portilla, J.; Castillo, J.-C. 2-Oxo-2*H*-chromen-7-yl 4-chlorobenzoate. *Molbank* **2021**, *2021*, M1279. [CrossRef]
10. Le, H.-T.-T.; Chau, Q.-C.; Duong, T.-H.; Tran, Q.-T.-P.; Pham, N.-K.-T.; Nguyen, T.-H.-T.; Nguyen, N.-H.; Sichaem, J. New Derivatives of Lupeol and Their Biological Activity. *Molbank* **2021**, *2021*, M1306. [CrossRef]

Short Note

2-(3,4-Dihydroxyphenyl)-4-(2-(4-nitrophenyl)hydrazono)-4H-chromene-3,5,7-triol

Alessandra Gianoncelli [1], Alberto Ongaro [1], Giuseppe Zagotto [2], Maurizio Memo [1] and Giovanni Ribaudo [1,*]

[1] Department of Molecular and Translational Medicine, University of Brescia, 25123 Brescia, Italy; alessandra.gianoncelli@unibs.it (A.G.); a.ongaro005@unibs.it (A.O.); maurizio.memo@unibs.it (M.M.)

[2] Department of Pharmaceutical and Pharmacological Sciences, University of Padova, 35131 Padova, Italy; giuseppe.zagotto@unipd.it

* Correspondence: giovanni.ribaudo@unibs.it; Tel.: +39-030-371-7419

Academic Editor: Fang-Rong Chang
Received: 10 June 2020; Accepted: 28 June 2020; Published: 29 June 2020

Abstract: On the basis of the knowledge from traditional herbal and folk medicine, flavonoids are among the most studied chemical classes of natural compounds for their potential activity as phosphodiesterase 5 (PDE5) inhibitors. We here describe the preparation of a semi-synthetic hydrazone derivative of quercetin, 2-(3,4-dihydroxyphenyl)-4-(2-(4-nitrophenyl)hydrazono)-4H-chromene-3,5,7-triol, that was obtained via a single-step modification of the natural compound. The product was characterized by NMR, mass spectrometry and HPLC. Preliminary molecular modeling studies suggest that this compound could efficiently interact with PDE5.

Keywords: quercetin; flavonoids; semi-synthetic; PDE; sildenafil; molecular modeling

---

1. Introduction

Phosphodiesterase (PDE) inhibitors contrast the degradation of 3′,5′-cyclic adenosine monophosphate (cAMP) and/or 3′,5′-cyclic guanosine monophosphate (cGMP), thus promoting several downstream effects. The inhibitors of PDE5 isoform, in particular, interfere with cGMP hydrolysis and induce smooth-muscle relaxation in specific tissues [1,2]. These compounds find clinical applications in the treatment of erectile dysfunction and pulmonary hypertension, and they are being studied as potential treatments against other diseases [3–5]. Repurposing of approved drugs is becoming an attractive strategy for identifying new applications for compounds with proved safety [6], and in this context PDE5 inhibitors are currently under investigation to contrast neurodegeneration [7,8], depression [9], diabetes [10] and rare pathologies such as Duchenne muscular dystrophy [11].

The development of synthetic PDE5 inhibitors, such as sildenafil (Figure 1A) and its analogues, has been paralleled by the exploration of the potential activity of natural compounds from traditional and folk medicine [12–14]. Flavonoids, in particular, have been known as PDE inhibitors for decades [15]. Natural glycosylated flavonoids and aglycones [16–19], as well as semi-synthetic flavones [20] and isoflavones [21–23], have been studied in silico, in vitro and in vivo for their inhibitory activity towards PDE5 and other isoforms.

The single-step derivatization procedure to obtain 2-(3,4-dihydroxyphenyl)-4-(2-(4-nitrophenyl)hydrazono)-4H-chromene-3,5,7-triol (**1**) from quercetin is here reported. This semi-synthetic compound was characterized by NMR, mass spectrometry and HPLC. Its interaction pattern with PDE5 was investigated in silico and compared to that of quercetin and sildenafil.

2. Results and Discussion

2.1. Chemistry

Quercetin was previously reported to possess inhibitory activity on several PDE isoforms [24], and the vasorelaxant effect of this natural flavonoid and of the corresponding metabolites was demonstrated to be due to the interference with the cGMP pathway [25]. Chan et al. investigated the effects of quercetin derivatives on PDE isoforms [26], and we previously explored the semi-synthetic derivatization of flavonoids to enhance their interaction with PDE5 in silico and in vitro [21,22].

We here report the preparation of a hydrazone derivative, 2-(3,4-dihydroxyphenyl)-4-(2-(4-nitrophenyl)hydrazono)-4*H*-chromene-3,5,7-triol (1), that was obtained via a single-step modification procedure from quercetin (Figure 1B). Rollas et al. recently discussed the biochemical relevance, in terms of reactivity and bioactivity aspects, of hydrazones [27]. Hydrazones are generally prepared from carbonyl compounds by reaction with an opportune hydrazine in acidic conditions [28,29]. Hydrochloric, acetic or 3-chloroperbenzoic acid are usually adopted for the synthesis of hydrazones, and the use of catalysts has been reported [30,31]. Compound 1 was synthesized by reacting quercetin with an excess of 4-nitrophenylhydrazine in a 1:1 mixture of acetic acid and ethanol. Following this procedure, the compound was isolated by filtration in a good yield (56%).

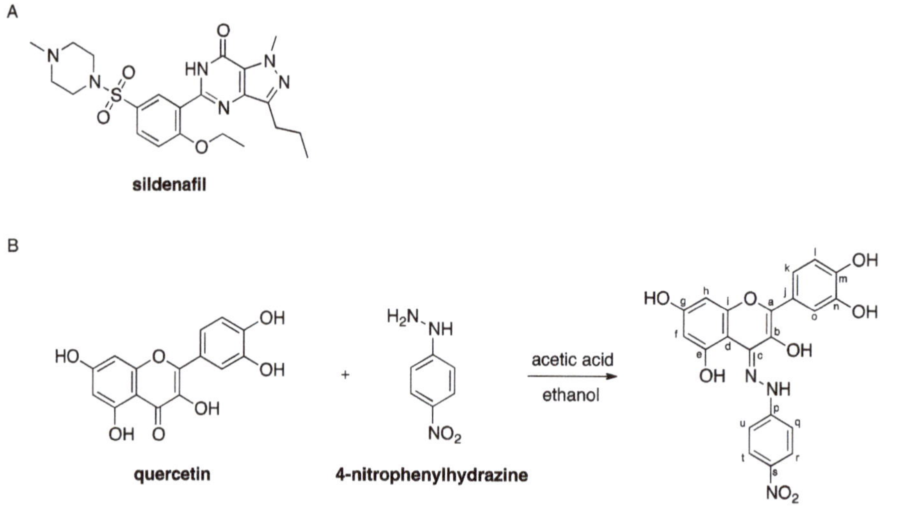

Figure 1. Chemical structure of sildenafil (**A**) and synthetic scheme for the preparation of compound **1** (**B**).

Compound **1** was characterized by NMR, mass spectrometry and HPLC (see Figures S1–S4 in the Supplementary material).

2.2. Molecular Modeling

The interaction of compound **1** with PDE5 was investigated in silico following a protocol reported previously [22]. For comparison, sildenafil and quercetin were also docked to the same 3D model and the predicted interaction patterns demonstrated a good co-localization of the ligands within the protein. The calculated binding energy value was particularly encouraging for compound **1** (−10.3 kcal/mol), exceeding that predicted for sildenafil (−9.7 kcal/mol) and quercetin (−9.5 kcal/mol) (Figure 2A). More in detail, docking experiments showed that the three compounds bind to the same region of the protein, consisting in the catalytic site (Figure 2B–D and Figures S5–S10 in the Supplementary materials). Most importantly, according to the predicted models, compound **1**, sildenafil and quercetin interact with the same group of residues in such a PDE5 domain. In particular, Ile778, Val782, Ala783, Leu804, Ile813, Met816, Gln817 and Phe820, which were previously reported to be relevant interacting residues for known PDE5 inhibitors [32], were highlighted within the < 5 Å region from the docked ligands (Figure 2).

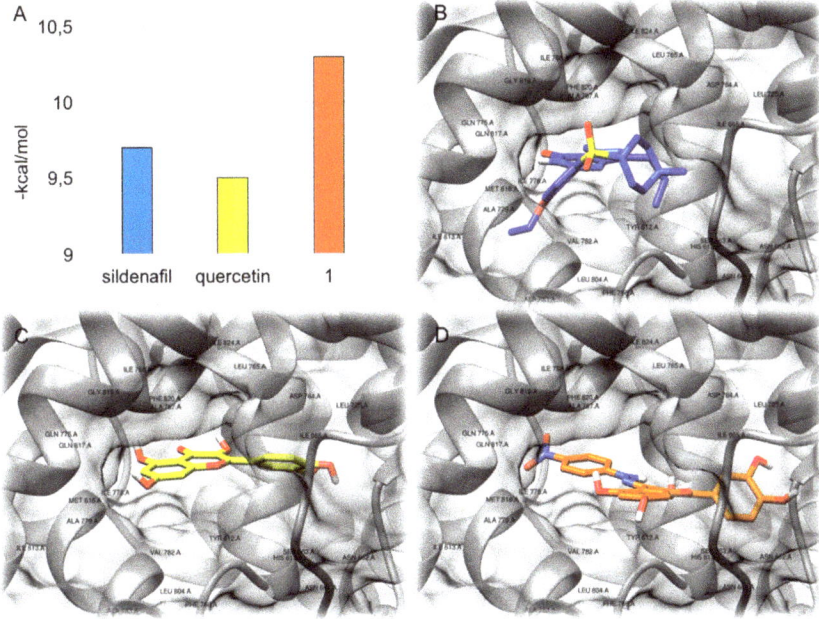

Figure 2. Results of the docking study for compound **1** to PDE5, in comparison with sildenafil and quercetin. (**A**): calculated binding energy values obtained from docking experiments. (**B**): docking pose of sildenafil. (**C**): docking pose of quercetin. (**D**): docking pose of compound **1**.

Furthermore, the stability of the complex predicted for compound **1** and PDE5 was assessed using molecular dynamics simulations [33]. The results show that the complex reached stability after 8 ns, and it was retained during the remaining simulation time (see Figures S11–S12 in the Supplementary materials)

3. Materials and Methods

3.1. Chemistry

3.1.1. General

Commercially available chemicals were purchased from Sigma–Aldrich (Saint Louis, MO, USA) and used as received, unless otherwise stated. ^{1}H and $^{13}C\{^{1}H\}$ NMR spectra were recorded on an Avance III 400 MHz spectrometer (Bruker, Billerica, MA, USA). All spectra were recorded at room temperature; the solvent for each spectrum is given in parentheses. Chemical shifts are reported in ppm and are relative to tetramethylsilane (TMS) internally referenced to the residual solvent peak. Datasets were edited with iNMR (Nucleomatica, Molfetta, Italy). The multiplicity of signals is reported as a singlet (s), doublet (d), triplet (t), quartet (q), multiplet (m), broad (b) or a combination of any of these. Mass spectra were recorded by direct infusion ESI on a Xevo G2-XS (Waters, Milford, MA, USA). The purity profile (96%) was assayed by HPLC using a Pro-Star system (Varian, Palo Alto, CA, USA) equipped with a 1706 UV–VIS detector (254 nm, Bio-rad, Hercules, CA, USA) and an C-18 column (5 µm, 4.6 × 250 mm, Agilent Technologies, Santa Clara, CA, USA). An appropriate ratio of water (A) and acetonitrile (B) was used as mobile phase with an overall flow rate of 1 mL/min; the general methods for the analyses are reported here: 0 min (95% A–5% B), 5 min (95% A–5% B), 25 min (5% A–95% B), 35 min (5% A–95% B) and 40 min (95% A–5% B).

3.1.2. Synthesis of 2-(3,4-dihydroxyphenyl)-4-(2-(4-nitrophenyl)hydrazono)-4*H*-chromene-3,5,7-triol (**1**)

A round-bottom flask was charged with quercetin (50.0 mg, 0.17 mmol) and ethanol (5 mL). A solution of 4-nitrophenylhydrazine (76.0 mg, 0.50 mmol) in acetic acid (5 mL) was added dropwise to this mixture and the reaction was refluxed under stirring for 6 h. After cooling to room temperature, the concentration of the solvent induced the formation of a precipitate. The solid, collected by filtration, was triturated using diethyl ether and the resulting product was isolated as a brown solid (42.0 mg). Yield: 56%, mp. 264–267 °C, HPLC r.t. 21.8 min. ^{1}H-NMR (DMSO-d_6, 400 MHz): δ_H, 12.48 (1H, bs, OH), 9.96 (1H, bs, OH), 8.99 (1H, bs, OH), 8.05 (2H, d, *J* 8.1 Hz, Hr and Ht), 7.67 (1H, d, *J* 1.8 Hz, Ho), 7.52 (1H, dd, *J* 8.0 Hz, *J* 1.8 Hz, Hk), 6.89 (1H, d, *J* 8.0 Hz, Hl), 6.73 (2H, d, *J* 8.1 Hz, Hq and Hu), 6.37 (1H, s, Hf of Hh), 6.15 (1H, s, Hh or Hf). ^{13}C {^{1}H}-NMR (DMSO-d_6, 100 MHz): δ_C 177.2, 176.3, 169.5, 164.3, 161.1, 159.2, 156.5, 155.4, 148.1, 147.2, 145.5, 138.4, 136.2, 126.3, 122.4, 120.4, 116.0, 115.5, 110.9. ESI-MS found 438.454 ($C_{21}H_{16}N_3O_8^+$. [M + H]$^+$), calc. 438.366.

3.2. Molecular Modeling

The structure of PDE5 was obtained from the RCSB Protein Data Bank (www.rcsb.org, PDB ID: 2H42). The target and ligands were prepared for the blind docking experiment which was performed using Autodock Vina (Molecular Graphics Laboratory, Department of Integrative Structural and Computational Biology, The Scripps Research Institute, La Jolla, CA, USA) [34]. Output data (energies, interaction patterns) were analyzed and scored using a UCSF Chimera molecular viewer [35], which was also used to produce the artworks. Molecular dynamics simulations were carried out using PlayMolecule (Accelera, Middlesex, UK) starting from the output model of docking experiments. A ligand was prepared by running a Parametrize function based on GAFF2 force field [36]. The complex was prepared for the simulation using ProteinPrepare and SystemBuilder functions, setting pH = 7.4, AMBER force field and default experiment parameters [37]. A simulation of 25 ns was carried out using SimpleRun, with default settings [38].

4. Conclusions

In this short note, we reported the preparation of 2-(3,4-dihydroxyphenyl)-4-(2-(4-nitrophenyl) hydrazono)-4*H*-chromene-3,5,7-triol (**1**), a semi-synthetic hydrazone derivative of quercetin that was obtained via a single-step approach. Preliminary in silico studies suggest that this compound could efficiently interact with the catalytic domain of PDE5 and that the effect on enzymatic inhibition of quercetin derivatives bearing this or other hydrazone substituents should be evaluated in vitro.

Supplementary Materials: The following are available online, Figures S1 and S2: NMR spectra, Figure S3: ESI-MS spectrum, Figure S4: HPLC profile, Figures S5–S10: docking studies, Figures S11 and S12: molecular dynamics simulations.

Author Contributions: Conceptualization, A.G. and G.R.; methodology, G.R.; software, A.O. and G.R.; investigation, A.O.; data curation, A.G.; writing—original draft preparation, A.G., A.O. and G.R.; writing—review and editing, M.M. and G.Z.; supervision, A.G. and G.R.; funding acquisition, A.G. and G.R. All authors have read and agreed to the published version of the manuscript.

Funding: This research was granted by University of Brescia.

Conflicts of Interest: The authors declare no conflict of interest.

References

1. Gur, S.; Kadowitz, P.J.; Serefoglu, E.C.; Hellstrom, W.J.G. PDE5 inhibitor treatment options for urologic and non-urologic indications: 2012 update. *Curr. Pharm. Des.* **2012**, *18*, 5590–5606. [CrossRef] [PubMed]

2. Andersson, K.-E. PDE5 inhibitors - pharmacology and clinical applications 20 years after sildenafil discovery. *Br. J. Pharmacol.* **2018**. [CrossRef]

3. Ribaudo, G.; Pagano, M.A.; Bova, S.; Zagotto, G. New Therapeutic Applications of Phosphodiesterase 5 Inhibitors (PDE5-Is). *Curr. Med. Chem.* **2016**, *23*, 1239–1249. [CrossRef] [PubMed]

4. Ahmed, N.S. Tadalafil: 15 years' journey in male erectile dysfunction and beyond. *Drug Dev. Res.* **2018**, ddr.21493. [CrossRef] [PubMed]

5. Nabavi, S.M.; Talarek, S.; Listos, J.; Nabavi, S.F.; Devi, K.P.; Roberto de Oliveira, M.; Tewari, D.; Argüelles, S.; Mehrzadi, S.; Hosseinzadeh, A.; et al. Phosphodiesterase inhibitors say NO to Alzheimer's disease. *Food Chem. Toxicol.* **2019**, *134*, 110822. [CrossRef] [PubMed]

6. Pushpakom, S.; Iorio, F.; Eyers, P.A.; Escott, K.J.; Hopper, S.; Wells, A.; Doig, A.; Guilliams, T.; Latimer, J.; McNamee, C.; et al. Drug repurposing: Progress, challenges and recommendations. *Nat. Rev. Drug Discov.* **2019**, *18*, 41–58. [CrossRef] [PubMed]

7. Zuccarello, E.; Acquarone, E.; Calcagno, E.; Argyrousi, E.K.; Deng, S.X.; Landry, D.W.; Arancio, O.; Fiorito, J. Development of novel phosphodiesterase 5 inhibitors for the therapy of Alzheimer's disease. *Biochem. Pharmacol.* **2020**, 113818. [CrossRef] [PubMed]

8. Ribaudo, G.; Ongaro, A.; Zagotto, G.; Memo, M.; Gianoncelli, A. Therapeutic Potential of Phosphodiesterase (PDE) Inhibitors Against Neurodegeneration: The Perspective of the Medicinal Chemist. *ACS Chem. Neurosci.* **2020**, acschemneuro.0c00244.

9. Duarte-Silva, E.; Filho, A.J.M.C.; Barichello, T.; Quevedo, J.; Macedo, D.; Peixoto, C. Phosphodiesterase-5 inhibitors: Shedding new light on the darkness of depression? *J. Affect. Disord.* **2020**, *264*, 138–149. [CrossRef]

10. Hackett, G. Should All Men with Type 2 Diabetes Be Routinely Prescribed a Phosphodiesterase Type 5 Inhibitor? *World J. Mens. Health* **2020**, *38*. [CrossRef]

11. Vitiello, L.; Tibaudo, L.; Pegoraro, E.; Bello, L.; Canton, M. Teaching an Old Molecule New Tricks: Drug Repositioning for Duchenne Muscular Dystrophy. *Int. J. Mol. Sci.* **2019**, *20*, 6053. [CrossRef] [PubMed]

12. Pavan, V.; Mucignat-Caretta, C.; Redaelli, M.; Ribaudo, G.; Zagotto, G. The Old Made New: Natural Compounds against Erectile Dysfunction. *Arch. Pharm. (Weinheim)* **2015**, *348*, 607–614. [CrossRef] [PubMed]

13. Ribaudo, G.; Zanforlin, E.; Canton, M.; Bova, S.; Zagotto, G. Preliminary studies of berberine and its semi-synthetic derivatives as a promising class of multi-target anti-parkinson agents. *Nat. Prod. Res.* **2018**, *32*, 1395–1401. [CrossRef]

14. Ribaudo, G.; Ongaro, A.; Zagotto, G. Natural Compounds Promoting Weight Loss: Mechanistic Insights from the Point of View of the Medicinal Chemist. *Nat. Prod. J.* **2019**, *9*, 78–85. [CrossRef]

15. Beretz, A.; Anton, R.; Stoclet, J.C. Flavonoid compounds are potent inhibitors of cyclic AMP phosphodiesterase. *Experientia* **1978**, *34*, 1054–1055. [CrossRef] [PubMed]

16. Sabphon, C.; Temkitthawon, P.; Ingkaninan, K.; Sawasdee, P. Phosphodiesterase Inhibitory Activity of the Flavonoids and Xanthones from Anaxagorea luzonensis. *Nat. Prod. Commun.* **2015**, *10*, 1934578X1501000222. [CrossRef]

17. Ko, W.-C.; Shih, C.-M.; Lai, Y.-H.; Chen, J.-H.; Huang, H.-L. Inhibitory effects of flavonoids on phosphodiesterase isozymes from guinea pig and their structure–activity relationships. *Biochem. Pharmacol.* **2004**, *68*, 2087–2094. [CrossRef]

18. Ribaudo, G.; Vendrame, T.; Bova, S. Isoflavones from *Maclura pomifera*: Structural elucidation and in silico evaluation of their interaction with PDE5. *Nat. Prod. Res.* **2017**, *31*, 1988–1994. [CrossRef] [PubMed]

19. Ongaro, A.; Zagotto, G.; Memo, M.; Gianoncelli, A.; Ribaudo, G. Natural phosphodiesterase 5 (PDE5) inhibitors: A computational approach. *Nat. Prod. Res.* **2019**, 1–6. [CrossRef] [PubMed]

20. Chau, Y.; Li, F.-S.; Levsh, O.; Weng, J.-K. Exploration of icariin analog structure space reveals key features driving potent inhibition of human phosphodiesterase-5. *PLoS ONE* **2019**, *14*, e0222803. [CrossRef]

21. Ribaudo, G.; Pagano, M.A.; Pavan, V.; Redaelli, M.; Zorzan, M.; Pezzani, R.; Mucignat-Caretta, C.; Vendrame, T.; Bova, S.; Zagotto, G. Semi-synthetic derivatives of natural isoflavones from Maclura pomifera as a novel class of PDE-5A inhibitors. *Fitoterapia* **2015**, *105*, 132–138. [CrossRef]

22. Ribaudo, G.; Ongaro, A.; Zagotto, G. 5-Hydroxy-3-(4-hydroxyphenyl)-8,8-dimethyl-6-(3-methylbut-2-enyl) pyrano [2,3-h]chromen-4-one. *Molbank* **2018**, *2018*, M1004. [CrossRef]
23. Ribaudo, G.; Coghi, P.; Zanforlin, E.; Law, B.Y.K.; Wu, Y.Y.J.; Han, Y.; Qiu, A.C.; Qu, Y.Q.; Zagotto, G.; Wong, V.K.W. Semi-synthetic isoflavones as BACE-1 inhibitors against Alzheimer's disease. *Bioorg. Chem.* **2019**, *87*, 474–483. [CrossRef] [PubMed]
24. Adefegha, S.A.; Oboh, G.; Fakunle, B.; Oyeleye, S.I.; Olasehinde, T.A. Quercetin, rutin, and their combinations modulate penile phosphodiesterase-5', arginase, acetylcholinesterase, and angiotensin-I-converting enzyme activities: A comparative study. *Comp. Clin. Path.* **2018**, *27*, 773–780. [CrossRef]
25. Suri, S.; Liu, X.; Rayment, S.; Hughes, D.; Kroon, P.; Needs, P.; Taylor, M.; Tribolo, S.; Wilson, V. Quercetin and its major metabolites selectively modulate cyclic GMP-dependent relaxations and associated tolerance in pig isolated coronary artery. *Br. J. Pharmacol.* **2010**, *159*, 566–575. [CrossRef] [PubMed]
26. Chan, A.L.F.; Huang, H.L.; Chien, H.C.; Chen, C.M.; Lin, C.N.; Ko, W.C. Inhibitory effects of quercetin derivatives on phosphodiesterase isozymes and high-affinity [3H]-rolipram binding in guinea pig tissues. *Invest. New Drugs* **2008**, *26*, 417–424. [CrossRef] [PubMed]
27. Rollas, S.; Küçükgüzel, Ş.G. Biological activities of hydrazone derivatives. *Molecules* **2007**, *12*, 1910–1939. [CrossRef]
28. Newkome, G.R.; Fishel, D.L. Synthesis of Simple Hydrazones of Carbonyl Compounds by an Exchange Reaction. *J. Org. Chem.* **1966**, *31*, 677–681. [CrossRef]
29. Hajipour, A.R.; Mohammadpoor-Baltork, I.; Bigdeli, M. A Convenient and Mild Procedure for the Synthesis of Hydrazones and Semicarbazones from Aldehydes or Ketones under Solvent-free Conditions. *J. Chem. Res.* **1999**, 570–571. [CrossRef]
30. Zhang, M.; Shang, Z.-R.; Li, X.-T.; Zhang, J.-N.; Wang, Y.; Li, K.; Li, Y.-Y.; Zhang, Z.-H. Simple and efficient approach for synthesis of hydrazones from carbonyl compounds and hydrazides catalyzed by meglumine. *Synth. Commun.* **2017**, *47*, 178–187. [CrossRef]
31. Ribaudo, G.; Scalabrin, M.; Pavan, V.; Fabris, D.; Zagotto, G. Constrained bisantrene derivatives as G-quadruplex binders. *Arkivoc* **2016**, *2016*, 145.
32. Cahill, K.B.; Quade, J.H.; Carleton, K.L.; Cote, R.H. Identification of Amino Acid Residues Responsible for the Selectivity of Tadalafil Binding to Two Closely Related Phosphodiesterases, PDE5 and PDE6. *J. Biol. Chem.* **2012**, *287*, 41406–41416. [CrossRef] [PubMed]
33. Salmaso, V.; Moro, S. Bridging Molecular Docking to Molecular Dynamics in Exploring Ligand-Protein Recognition Process: An Overview. *Front. Pharmacol.* **2018**, *9*. [CrossRef] [PubMed]
34. Trott, O.; Olson, A.J. AutoDock Vina: Improving the speed and accuracy of docking with a new scoring function, efficient optimization, and multithreading. *J. Comput. Chem.* **2010**, *31*, 455–461. [CrossRef] [PubMed]
35. Pettersen, E.F.; Goddard, T.D.; Huang, C.C.; Couch, G.S.; Greenblatt, D.M.; Meng, E.C.; Ferrin, T.E. UCSF Chimera—A visualization system for exploratory research and analysis. *J. Comput. Chem.* **2004**, *25*, 1605–1612. [CrossRef]
36. Galvelis, R.; Doerr, S.; Damas, J.M.; Harvey, M.J.; De Fabritiis, G. A Scalable Molecular Force Field Parameterization Method Based on Density Functional Theory and Quantum-Level Machine Learning. *J. Chem. Inf. Model.* **2019**, *59*, 3485–3493. [CrossRef]
37. Martínez-Rosell, G.; Giorgino, T.; De Fabritiis, G. PlayMolecule ProteinPrepare: A Web Application for Protein Preparation for Molecular Dynamics Simulations. *J. Chem. Inf. Model.* **2017**, *57*, 1511–1516. [CrossRef]
38. Doerr, S.; Harvey, M.J.; Noé, F.; De Fabritiis, G. HTMD: High-Throughput Molecular Dynamics for Molecular Discovery. *J. Chem. Theory Comput.* **2016**, *12*, 1845–1852. [CrossRef]

Sample Availability: Samples of the compounds of compound **1** are available from the authors.

Short Note

5,5'-Oxybis(1,3,7-trihydroxy-9*H*-xanthen-9-one): A New Xanthone from the Stem Bark of *Garcinia porrecta* (Clusiaceae)

Ayu N. Safitri [1]**, Nurlelasari** [1]**, Tri Mayanti** [1]**, Darwati** [1] **and Unang Supratman** [1,2,]*

[1] Departement of Chemistry, Faculty of Mathematics and Natural Sciences, Universitas Padjadjaran, Jatinangor 45363, West Java, Indonesia; ayu15003@mail.unpad.ac.id (A.N.S.); nurlelasari@unpad.ac.id (N.); t.mayanti@yahoo.co.id (T.M.); darwatititi@gmail.com (D.)
[2] Central Laboratory, Universitas Padjadjaran, Jatinangor 45363, West Java, Indonesia
*   Correspondence: unang.supratman@unpad.ac.id; Tel.: +62-22-779-4391

Received: 15 July 2020; Accepted: 7 August 2020; Published: 12 August 2020

Abstract: A new polyoxygenated dimer-type xanthone, namely 5,5'-oxybis(1,3,7-trihydroxy-9*H*-xanthen-9-one (**1**), has been isolated from the stem bark of *Garcinia porrecta*. The structure of **1** was determined based on spectroscopic data, including 1D and 2D-NMR as well as high resolution mass spectroscopy analysis.

Keywords: *Garcinia porrecta*; Clusiaceae; xanthone

---

1. Introduction

The famous *Garcinia* genus, representing a major source of triterpenes, flavonoids, xanthones, and phloroglucinols which have pharmacological activities as antioxidants, antibacterial, antiviral, anti-HIV, and significant anticancer activity [1].

The genus *Garcinia* belongs to the Clusiaceae family, which consists of more than 400 species widely distributed in the Polinesia mainland, India, Indochina, Indonesia, West and Central Africa, and Brazil [2]. Indonesia is known as one of the countries rich in diversity of *Garcinia*, there are 64 species of *Garcinia* scattered across several islands in Indonesia [3]. Various parts of *Garcinia* plants have been used in traditional medicine for the treatment of sprue (mouth ulcer), diarrhea, dysentery and skin disease [4]. Investigations into biologically active compounds from Indonesia *Garcinia* plants have resulted in some bioactive compounds being isolated from *G. mangostana* [5–7], *G. celebica* [8,9] and *G. cowa* [10]. Previous investigation on the stem bark of *G. porrecta* had led to the isolation of dulxanthone E–G, which showed strong cytotoxic activity against murine leukemia L1210 cells [6]. In this paper, we reported the isolation and structure elucidation of new polyoxygenated dimer-type xanthone, 5,5'-Oxybis(1,3,7-trihydroxy-9*H*-xanthen-9-one) (**1**) (Figure 1).

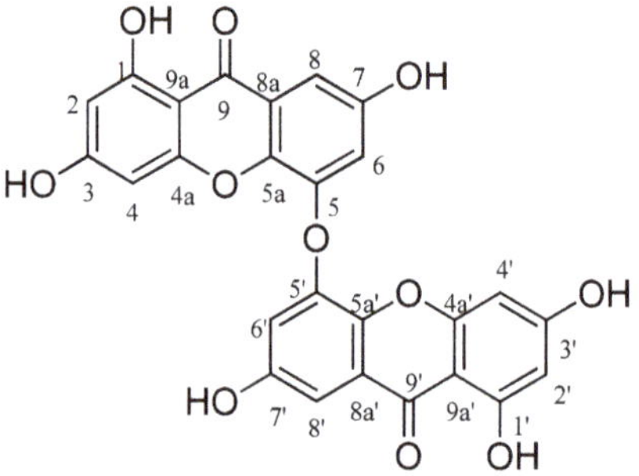

Figure 1. Chemical structure of compound **1**.

2. Results

Extraction and Isolation

The chopped dried stem bark of *G. porrecta* (2 Kg) was macerated at room temperature with *n*-hexane (5 × 2 L), ethyl acetate (5 × 2 L), and methanol (5 × 2 L). The solvents were removed by a rotary evaporator to give a crude *n*-hexane extract (21 g), ethyl acetate (12.5 g), and methanol (25 g). The ethyl acetate extract (12.5 g) was fractionated by vacuum liquid chromatography on silica gel using a gradient of *n*-hexane-ethyl acetate-methanol solvent to give eight fractions (A–H). Fraction E (1.93 g) was separated with silica gel column chromatography using *n*-hexane:methylene chloride:acetone (5:3:2) as the solvent system to give nine subfractions (E1–E9). Subfraction E8 (140.7 mg) was purified by column chromatography on RP-18 silica using 10% gradient MeOH:H$_2$O to give **1** (30.6 mg).

5,5'-Oxybis(1,3,7-trihydroxy-9*H*-xanthen-9-one) (**1**), yellow amorphous powder, $[\alpha]^{20}_D$ +12.4 (*c* 0.1, MeOH); UV (MeOH) λ_{max}: 322 and 262 nm; HR-TOFMS *m/z* 503.0667 [M + H]$^+$ (calcld. for C$_{26}$H$_{15}$O$_{11}$, 503.0614); IR (KBr) ν_{max}: 3412, 2962, 1755, 1484, 1174 cm^{-1}; ^{1}H-NMR (acetone-*d$_6$*, 600 MHz) δ_H: 6.2 (1H, s, H-2, H-2'), 6.3 (1H, s, H-4, H-4'), 6.9 (1H, s, H-8, H-8'), 7.5 (1H, s, H-6, H-6'), 13.2 (1H, s, OH-1); ^{13}C-NMR and DEPT-135 (acetone-*d$_6$*, 150 MHz), δ_c: 179.6 (C-9), 179.5 (C-9'), 164.8 (C-3'), 164.7 (C-1'), 163.5 (C-1), 163.2 (C-3), 157.9 (C-4a, C-4a'), 153.5 (C-5a, C-5a'), 151.6 (C-7, C-7'), 143.3 (C-5, C-5'), 122.7 (C-8a'), 112.8 (C-8a), 108.2 (C-6, C-6'), 102.5 (C-8, C-8'), 102.2 (C-9a), 102.1 (C-9a'), 97.7 (C-2), 97.6 (C-2'), 93.5 (C-4), 93.4 (C-4').

3. Discussion

Compound **1** was isolated as a yellow amorphous powder. The UV spectrum showed absorption bands at λ_{max} 322 and 262 nm attributable to a conjugated system [11,12]. Its molecular composition was established to be C$_{26}$H$_{14}$O$_{11}$ with twenty degrees of unsaturation from HR-TOFMS *m/z* 503.0667 [M + H]$^+$, calculated for C$_{26}$H$_{15}$O$_{11}$ (*m/z* 503.0614) and NMR spectral data (Table 1). The IR spectrum exhibited bands at ν_{max} 3412 cm^{-1} (hydroxyl), 2962 cm^{-1} (C-H stretching of aliphatic) and 1755 cm^{-1} (carbonyl).

The ^{13}C-NMR spectrum demonstrated the presence of a total of 26 carbon signals, which were classified by their chemical shifts, DEPT, and HSQC spectra (Figures S3 and S4) as eight sp^2 methine carbons, two carbonyl carbon at δ_C 179.53 and 179.53, 16 sp^2 quaternary carbons (including two sp^2 carbons and 14 sp^2 oxygenated carbons). These functionalities accounted for 14 out of the total 20 degrees of unsaturations. The remaining of six degrees of unsaturation were consistent with six cyclics of bixanthones [13,14].

The ^{1}H-NMR and HSQC spectra of **1** (Figures S1 and S4), showed proton signals indicative of a tetrasubstituted aromatic group δ_H 6.2 (2H, s, H-2, H-2'), 6.3 (2H, s, H-4, H-4'), 6.9 (2H, s, H-8, H-8') and 7.5 (2H, s, H-6, H-6') and showed a hydroxyl proton at δ_H 13.2 (1H, s, OH-1). Low shimming quality could explain the missing splitting of H-2/H-2', H-4/H-4'/H-6/H-6', and H-8/H-8' signals in the ^{1}H-NMR spectrum.

A comparison of the NMR data of **1** with 1,3,7-trihydroxyxanthone, gentisein, isolated from *Gentiana lutea* [14] indicated that the structure of compound **1** is very similar to gentisein. The main difference was the presence of dimer skeleton at C-5. The substitution of the xanthone skeleton was determined by HSQC and HMBC spectra (Figures S4 and S5). The HMBC correlations (Figure 2) from H-6 (δ_H 108.2) with C-5a (δ_C 153.5), C-5′ (δ_C 143.3), C-5 (δ_C 143.3) and C-7 (δ_C 151.6), and of H-6′ (δ_H 108.2) with C-5a′ (δ_C 153.5), C-5 (δ_C 143.3), (δ_C 143.3) and C-7′ (δ_C 151.6) suggested that the substituent of dimer xanthone with the xanthone at C-5 or C-5′. The hydroxyl group was located at C-1 based on HMBC correlations from OH-1 (δ_H 13.2) to C-1 (δ_C 163.5), C-2 (δ_C 97.7) and C-9a (δ_C 102.2) (Figure 2 and Figure S5). Therefore, the structure of **1** was assigned as 5,5′-Oxybis(1,3,7-trihydroxy-9*H*-xanthen-9-one).

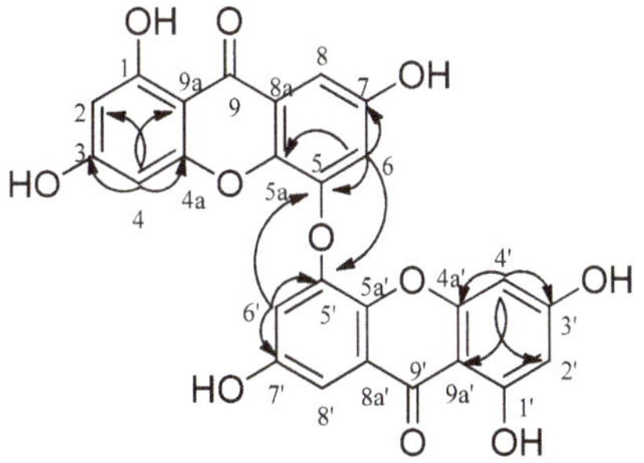

Figure 2. Selected HMBC correlations for **1**.

Table 1. NMR data of compound **1** and gentisein acetone-d_6.

Position	1		Gentisein [14]	
	δ_H [$\sum$H, mult., *J* (Hz)]	δ_C (mult.)	δ_H [$\sum$H, mult., *J* (Hz)]	δ_C
1		163.53 (s)		162.5
2	6.22 (1H, s)	97.69 (d)	6.2 (1H, d, 1.7)	97.8
3		163.22 (s)		165.5
4	6.37 (1H, s)	93.55 (d)	6.3 (1H, d, 1.7)	93.8
4a		157.97 (s)		157.6
5		143.38 (s)	7.4 (1H, d, 9.0)	119.1
5a		153.58 (s)		149.1
6	7.53 (1H, s)	108.21 (d)	7.3 (1H, dd, 9.0, 2.8)	124.5
7		151.67 (s)		153.8
8	6.91 (1H, s)	102.51 (d)	7.4 (1H, d, 2.8)	108.0
8a		112.81 (s)		120.5
9		179.60 (s)		179.7
9a		102.19 (s)		102.0
1′		164.74 (s)		
2′	6.22 (1H, s)	97.63 (d)		
3′		164.79 (s)		
4′	6.37 (1H, s)	93.48 (d)		
4a′		157.96 (s)		
5′		143.38 (s)		
5a′		153.55 (s)		
6′	7.53 (1H, s)	108.2 (d)		
7′		151.65 (s)		
8′	6.91 (1H, s)	102.51 (d)		
8a′		122.77 (s)		
9′		179.53 (s)		
9a′		102.16 (s)		
1,1′-OH	13.21 (1H, s)			

4. Materials and Methods

4.1. General Experimental Procedures

UV spectra were recorded on Vilber Lourmat UV/VIS spectrophotometer. Mass spectra were measured with a Waters Xevo QTOFMS instrument (Waters, Milford, MA, USA). IR spectra were measured on a One Perkin Elmer infrared-100. NMR data were recorded on a Bruker Avance-600 spectrometer at 600 MHz for ^{1}H and 150 MHz for ^{13}C using Tetramethylsilane (TMS) as an internal standard (Billerica, MA, USA). Chromatographic separations were carried out on silica gel G60 (0.063–0.200 mm) (Merck, Darmstadt, Germany), RP18 (0.04–0.063 mm) (Merck, Darmstadt, Germany). Precoated silica gel GF$_{254}$ plates (0.25 mm, Merck, Darmstadt, Germany) were used for Thin Layer Chromatography (TLC), and detection was achieved by spraying with 5% AlCl$_3$ in ethanol, followed by heating.

4.2. Plant Material

The stem bark of *G. porrecta* was collected from Bogor Botanical Garden, Bogor, Indonesia in April 2018. The plant was identified and deposited in the Herbarium Bogoriense (No. IV.K.78a), Center of Biological Research and Development, National Institute of Science, Bogor, Indonesia.

5. Conclusions

A new polyoxygenated dimer-type xanthone, namely 5,5′-Oxybis(1,3,7-trihydroxy-9*H*-xanthen-9-one) (**1**), was isolated from the stem bark of *G. porrecta*, belonging to Clusiaceae family. This polyoxygenated dimer-type xanthone was found in the *Garcinia* genus for the first time.

Supplementary Materials: The following are available online, Figure S1. ^{1}H-NMR spectrum of **1** (600 MHz in acetone-d_6), Figure S2. ^{13}C-NMR spectrum of **1** (150 MHz in acetone-d_6), Figure S3. DEPT-135° spectrum of **1** (150 MHz in acetone-d_6), Figure S4. HSQC Spectrum of **1**, Figure S5. HMBC spectrum of **1**, Figure S6. Infrared Spectrum of **1** (in KBr), Figure S7. HR-TOF-MS Spectrum of **1**, Figure S8. TLC Profile of **1**.

Author Contributions: Conceptualization, D., N., U.S.; Data curation, T.M.; N.; A.N.S.; Formal Analysis, N.; Investigation, A.N.S.; Methodology, A.N.S., N., T.M.; Supervision, D., U.S. All authors have read and agreed to the published version of the manuscript.

Funding: This research was funded by the Directorate General of Scientific Resources, Technology, and Higher Education, Ministry of Research, Technology, and Higher Education, Indonesia (PDUPT, number, 2788/UN6.D/LT/2019, by D.S.).

Acknowledgments: Authors also thank Mohamad Nurul Azmi from Universiti Sains Malaysia for the assistance of NMR measurements, Kansi Haikal at the Central Laboratory, Universitas Padjadjaran for QTOFMS Measurements, Suharto from Bogor Botanical Garden, Bogor for the plant sample.

Conflicts of Interest: The authors declare no conflict of interest.

References

1. Sari, A.C.; Elya, B.; Katrin. Antioxidant activity and lipoxygenase enzyme inhibition assay with total flavonoid assay of *Garcinia porrecta* laness. stem bark extracts. *Pharm. J.* **2017**, *9*, 257–266. [CrossRef]
2. Brito, L.C.; Berenger, A.L.R.; Figueiredo, M.R. An overview of anticancer activity of *garcinia* and *hypericum*. *Food Chem. Toxicol.* **2017**, *109*, 847–862. [CrossRef] [PubMed]
3. Kardono, L.B.S.; Hanafi, M.; Sherley, G.; Kosela, S.; Harrison, L.J. Bioactive constituents of *Garcinia porrecta* and *Garcinia parvifolia* grown in Indonesia. *J. Biol. Sci.* **2006**, *9*, 483–486.
4. Putri, I.P. Effectivity of xanthone of mangosteen (*Garcinia mangostana*) rind as anticancer. *J. Major.* **2015**, *4*, 33–38.
5. Chin, Y.W.; Kinghorn, A.D. Structural characterization, biological effects, and synthetic studies on xanthones from mangosteen (*Garcinia mangostana*), a popular botanical dietary supplement. *Mini Rev. Org. Chem.* **2008**, *5*, 355–364. [CrossRef] [PubMed]
6. Suksamrarn, S.; Komutiban, O.; Ratananukul, P.; Chimnoi, N.; Lartpornmatulee, N.; Suksamrarn, A. Cytotoxic prenylated xanthones from the young fruit of *Garcinia mangostana*. *Chem. Pharm. Bull.* **2006**, *54*, 301–305. [CrossRef] [PubMed]

7. Ibrahim, S.R.M.; Abdallah, H.M.; El-Halawany, A.M.; Radwan, M.F.; Shehata, I.A.; Al-Harshany, E.; Zayed, M.F.; Mohamed, G.A. Garcixanthones b an c, new xanthones from the pericarps of *Garcinia mangostana* and their cytotoxic activity. *Phytochem. Lett.* **2018**, *25*, 12–16. [CrossRef]

8. Subarnas, A.; Diantinil, A.; Abdulah, R.; Zuhrotun, A.; Nugraha, P.A.; Hadisaputri, Y.E.; Puspitasari, I.M.; Yamazaki, C.; Kuwano, H.; Koyama, H. Apoptosis-mediated antiproliferative activity of friedolanostane triterpenoid isolated from the leaves of *Garcinia celebica* against MCF7 human breast cancer cell lines. *Biomed. Rep.* **2016**, *4*, 79–82. [CrossRef] [PubMed]

9. Bui, T.Q.; Bui, A.; Nguyen, K.T.; Nguyen, V.T.; Trinh, B.T.D.; Nguyen, L.D. A depsidone and six triterpenoids from the bark of *Garcinia celebica*. *Tetrahedron Lett.* **2016**, *57*, 2524–2529. [CrossRef]

10. Ritthiwigrom, T.; Laphookhieo, S.; Pyne, S.G. Chemical constituents and biological activities of *Garcinia cowa* Roxb. *J. Sci. Technol.* **2013**, *7*, 212–231.

11. Shiono, Y.; Miyazani, N.; Murayama, T.; Koseki, T.; Harizon; Katja, D.G.; Supratman, U.; Nakata, J.; Kakihara, Y.; Saeki, M.; et al. GSK-3β inhibitory activities of novel dichloresorcinol derivatives from *Cosmospora vilior* isolated from a mangrove plant. *Phytochem. Lett.* **2016**, *18*, 122–127. [CrossRef]

12. Susilawati, Y.; Nugraha, R.; Muhtadi, A.; Soetardjo, S.; Supratman, U. (S)-2-Methyl-2-(4-methylpent-3-enyl)-6-(propan-2-ylidene)-3,4,6,7-tetra-hydropyrano [4,3-g]chromen-9(2H)-one. *Molbank* **2015**, *2015*, M855. [CrossRef]

13. Prakash, O.; Sing, R.; Kumar, S.; Srivastava, S.; Ved, A. *Gentiana lutea* Linn. (Yellow Gentian): A comprehensive review. *J. Ayurdevic Herb. Med.* **2017**, *3*, 175–181.

14. Ebrahim, W.; El-Neketi, M.; Lewald, L.I.; Orfali, R.S.; Lin, W.; Rehberg, N.; Kalscheuer, R.; Daletos, G.; Proksch, P. Metabolites from the fungal endophyte Aspergillus austroafricanus in axenic culture and in fungul-bacterial mixed cultures. *J. Nat. Prod.* **2016**, *79*, 914–922. [CrossRef] [PubMed]

Short Note

3-Hydroxy-8,14-secogammacera-7,14-dien-21-one: A New Onoceranoid Triterpenes from *Lansium domesticum* Corr. cv *kokossan*

Zulfikar [1], **Nurul Kamila Putri** [1], **Sofa Fajriah** [2], **Muhammad Yusuf** [1], **Rani Maharani** [1], **Jamaludin Al Anshori** [1], **Unang Supratman** [1,3] **and Tri Mayanti** [1,*]

[1] Department of Chemistry, Faculty of Mathematics and Natural Sciences, Universitas Padjadjaran, Jatinangor 45363, West Java, Indonesia; zulfikar15002@mail.unpad.ac.id (Z.); kpnurul@gmail.com (N.K.P.); m.yusuf@unpad.ac.id (M.Y.); r.maharani@unpad.ac.id (R.M.); jamaludin.al.anshori@unpad.ac.id (J.A.A.); unang.supratman@unpad.ac.id (U.S.)

[2] Research Center for Chemistry, Indonesian Science Institute, Serpong 15314, Banten, Indonesia; sofa002@lipi.go.id

[3] Central Laboratory, Universitas Padjadjaran, Jatinangor 45363, West Java, Indonesia

[*] Correspondence: t.mayanti@unpad.ac.id; Tel.: +62-0813-201-02633

Received: 24 August 2020; Accepted: 10 September 2020; Published: 30 September 2020

Abstract: A new onoceranoid triterpenes, namely 3-hydroxy-8,14-secogammacera-7,14-dien-21-one (**1**), has been isolated from the fruit peels of *Lansium domesticum* Corr. cv *kokossan*. The structure of **1** was determined on the basis of spectroscopic data including infrared, 1D and 2D-NMR, as well as high resolution mass spectroscopy analysis. Compound **1** showed a weak activity against MCF-7 breast cancer cell lines.

Keywords: *Lansium domesticum*; Meliaceae; MCF-7; triterpene onoceranoid

1. Introduction

Lansium domesticum Corr. (Meliaceae) is a popular plant that widely grows in southeastern Asia [1,2]. *L. domesticum* Corr. cv *kokossan* (Meliaceae) is a higher tree growing up to 30 m in height, commonly called "kokosan" in Indonesia [3]. Several bioactive triterpenoids have been isolated from *L. domesticum* [4–9], of which some have shown potential anticancer [10], antibacterial [11], insecticides [12], antimalarial [13], antimutagenic [14,15], and antifeedant activities [16]. Previously, we isolated six triterpenes from the seeds and bark of *L. domesticum* Corr. cv *kokossan* [2,16]. In this paper, we report the isolation and structural elucidation of the new onoceranoid triterpenes **1** (Figure 1), along with its cytotoxic activity against MCF-7 breast cancer lines.

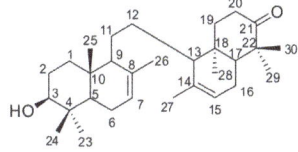

Figure 1. Chemical Structure of compound **1**.

2. Results

Extraction and Isolation

The dried fruit peel (1.7 kg) was extracted with *n*-hexane, ethyl acetate, and methanol three times, respectively, for 24 h, at room temperature. After solvent removal under reduced pressure, we obtained the *n*-hexane (86 g), ethyl acetate (110 g), and methanol (75 g) crude extracts. A portion of *n*-hexane (60 g) was subjected to vacuum liquid chromatography over silica gel using a 10% gradient of *n*-hexane/EtOAc (10:0–0:10) to afford eight fractions (A–H). Fraction D (6.7 g) was subjected to vacuum liquid chromatography, eluted by a 1% gradient of *n*-hexane/EtOAc (100:0–85:15) to afford five combined fractions (D1–D5). Fraction D4 was then subjected to silica gel column chromatography, eluted with *n*-hexane:acetone:methanol (7:2.5:0.5), resulting in four fractions (D4A–D4D). Compound **1** was obtained from fraction D4B as a white crystal (0.27 g).

3-Hydroxy-8,14-secogammacera-7,14-dien-21-one (**1**), white crystal, m.p. 153–155 °C, $[\alpha]^{20}_D - 3.0°$ (c 0.1, MeOH), HR-TOF-MS *m/z* 463.3569 [M + Na]$^+$ (calcd. for $C_{30}H_{48}O_2Na$, *m/z* 463.3552); IR (KBr) ν_{max} (cm^{-1}): 3533, 2932, 1700, 1456; ^{1}H-NMR and ^{13}C-NMR showed in the Table 1. Compound **1** was evaluated for its cytotoxic acitivity against MCF-7 breast cancer cell line, and compared to doxorubicin (35.7 µM) as a positive control. Compound **1** exhibited weak activity against MCF-7 with an IC$_{50}$ value of 717.5 µM.

Table 1. ^{13}C and ^{1}H NMR Spectral Data of Compounds **1** in CDCl$_3$.

Position	δ_C ppm	δ_H ppm (Int, mult., J = Hz)
1	37.5	1.14; 1.86 (2H, *m*)
2	27.4	1.65 (2H, *m*)
3	79.1	3.29 (1H, *dd*, 11.3; 4.3)
4	38.7	-
5	51.5	1.62 (1H, *m*)
6	29.8	1.35; 1.50 (2H, *m*)
7	121.7	5.40 (1H, *brs*)
8	135.4	-
9	55.3	1.66 (1H, *m*)
10	36.5	-
11	24.1	1.99; 2.10 (2H, *m*)
12	23.5	1.98; 2.10 (2H, *m*)
13	56.0	1.59 (1H, *m*)
14	134.9	-
15	122.3	5.40 (1H, *brs*)
16	29.9	1.35; 1.50 (2H, *m*)
17	49.6	1.20 (1H, *m*)
18	36.5	-
19	38.4	1.51; 2.15 (2H, *m*)
20	34.7	2.29; 2.77 (2H, *m*)
21	217.0	-
22	47.5	-
23	27.9	1.00 (3H, *s*)
24	15.1	0.87 (3H, *s*)
25	13.3	0.99 (3H, *s*)
26	22.3	1.75 (3H, *s*)
27	22.4	1.72 (3H, *s*)
28	13.6	0.76 (3H, *s*)
29	22.1	1.12 (3H, *s*)
30	25.0	1.08 (3H, *s*)

3. Discussion

The molecular composition of **1** was proposed as $C_{30}H_{48}O_2$ with seven degrees of unsaturation, based on HR-TOF-MS and NMR spectral data. Mass spectra showed molecular ion peak at *m/z* 463.3569 $[M + Na]^+$, with calculated *m/z* 463.3552 for $C_{30}H_{48}O_2Na$. The IR spectrum exhibited bands at v_{max} (cm^{-1}) 3533 (hydroxy), 2932 (C-H stretching of aliphatic), 1700 (ketone), and 1456 (*gem* dimethyl).

The ^{13}C NMR data (Table 1) with DEPT and HSQC experiments (Figures S2–S4) revealed the presence of total of 30 carbon signals, which were classified as eight methyls, eight methylenes, seven methines (two olefinic and one oxygenated), and seven quaternary carbons (two olefinic and one carbonyl). The two trisubstituted double bonds and the carbonyl in a system with seven degrees of unsaturation suggested that **1** possess a tetracyclic structure. Previous studies supported that evidence, indicating that compound **1** has an onoceranoid-type triterpenoid [3,14].

The ^1H NMR spectrum of **1** (Figure S1) displayed the presence of eight methyl groups at δ_H (ppm) 1.00 (3H, Me-23), 0.87 (3H, Me-24), 0.99 (3H, Me-25), 1.75 (3H, Me-26), 1.72 (3H, Me-27), 0.76 (3H, Me-28), 1.12 (3H, Me-29), and 1.08 (3H, Me-30). Two olefinic protons (H-7 and H-15) were observed at δ_H (ppm) 5.45 and 5.42 (each 1H, brs), together with two vinyl methyls proton resonating at δ_H (ppm) 1.75 (3H, s, H-26) and δ_H 1.72 (3H, s, H-27), indicating two trisubstituted double bonds of **1**. One oxygenated methine signal was also observed at δ_H 3.29 ppm (1H, dd, H-3).

The structure of **1** was further defined by ^1H-^1H COSY (correlated spectroscopy) and HMBC (heteronuclear multiple bond connectivity) spectra (Figure 2, Figures S5 and S6). The ^1H-^1H COSY spectra showed couplings between H-1 (δ_H 1.14)/H-2 (δ_H 1.65), H-2 (δ_H 1.65)/H-3 (δ_H 3.29), H-9 (δ_H 1.66)/H-11 (δ_H 2.10), H-11 (δ_H 2.10)/H-12 (δ_H 1.98), H-12 (δ_H 1.98)/H-13 (δ_H 1.59), and H-19 (δ_H 1.51)/H-20 (δ_H 2.77), supporting the presence of a onoceranoid-type triterpenoid structure. The oxygenated methine bearing a hydroxy group was located at C-3 by the HMBC correlations of H_2-1 (δ_H 1.86), H_3-23 (δ_H 1.00) and H_3-24 (δ_H 0.87) to C-3 (δ_C 79.1). The $\Delta^{7,8}$ and $\Delta^{14,15}$ double bonds were assigned by the HMBC correlations from H_3-26 (δ_H 1.75) to C-7 (δ_C 121.7), C-8 (δ_C 135.4), C-9 (δ_C 55.3); and H_3-27(δ_H 1.72) to C-13 (δ_C 56.0), C-14 (δ_C 134.9), C-15 (δ_C 122.3). The ketone group was attached to C-21 by the HMBC correlations of H_2-20 (δ_H 2.77) and H_3-29 (δ_H 1.12) to C-21 (δ_C 217.0). This analysis indicated that compound **1** was similar to 3β-hydroxyonocera-8(26),14-dien-21-one [5], uniquely differing on the double bond position in the B ring. Based on the coupling constants of H-3 (dd, J = 11.4; 4.3 Hz), the configuration of the hydroxyl group at C-3 was indicated in the β-equatorial position [5,17]. From the analyses, compound **1** was determined as 3-hydroxy-8,14-secogammacera-7,14-dien-21-one.

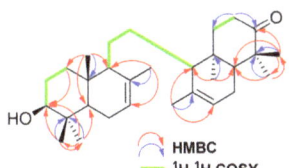

Figure 2. ^1H-^1H COSY and HMBC correlations of compound **1**.

4. Materials and Methods

4.1. General Experimental Procedures

Mass spectrum was recorded on a waters Xevo QTOFMS (Waters, Milford, MA, USA). IR spectrum was measured on a One Perkin Elmer infrared-100 (Waltham, MA, USA) in KBr. NMR data were recorded on a Brucker spectrometer (Billerica, MA, USA) at 400 MHz for ^1H and 120 MHz for ^{13}C using TMS as an internal standard. Chromatographic separations were carried out on silica gel G60 (Merck, Darmstadt, Germany) and RP18 (Merck). TLC plates were precoated with silica gel GF254 (Merck, 0.25 mm) and detection was achieved by spraying with 10% (*v/v*) H_2SO_4 in ethanol, followed by heating.

4.2. Plant Material

The fruit peels of *L. domesticum* Corr. cv *kokossan* (Meliaceae) was collected from Cililin, West Java, Indonesia in April 2018. The plant was identified and deposited in The Herbarium of Department of Biology, Faculty of Mathematics and Natural Sciences, Universitas Padjadjaran, Indonesia (Identification Number: 195/HB/08/2018).

4.3. Cytotoxic Bioassay

Cytotoxicity of the compound against MCF-7 human breast cancer cells was measured using MTT (*Methyl Thiazoldiphenyl-Tetrazoliumbromide*) assay. Stock culture was grown in flasks, containing Roswell Park Memorial Institute (RPMI) medium, respectively supplemented with 10% (*v/v*) feta bovine serum (FBS) and 1% (*v/v*) penicillin-streptomicin as an antibiotic. The culture was incubated at 37 °C for 24 h. The medium was changed, and tumor cells were detached and seeded in 96-well microliter plates. After, 24 h, compounds were added to the wells. After 48 h, cell viability was determined by measuring the metabolic conversion of yellow salt or 3-(4,5-dimethyltiazol-2-yl)-2,5-diphenyltetrazolium bromide to its insoluble formazan, which has a purple color product resulting from reduction in viable cells. Insoluble formazan was diluted with DMSO. The MTT assay results were read using spectrophotometer UV at 450 nm. Compound **1** was evaluated at eight concentrations (15.9; 34.0; 70.3; 140.7; 283.6; 567.3; 1134.6; 2269.1 μM) in 100% DMSO with the final concentration of DMSO of 2.5% (*v/v*) in each well. Doxorubicin was used as a positive control. IC_{50} values were calculated by the linear regression method using Microsoft Excel software.

5. Conclusions

A new onoceranoid triterpene, namely 3-hydroxy-8,14-secogammacera-7,14-dien-21-one (**1**), was isolated from fruit peels of *L. domesticum* Corr. cv *kokossan* (Meliaceae). This compound exhibited weak cytotoxic activity against MCF-7 human breast cancer cell lines with IC_{50} value of 717.5 μM.

Supplementary Materials: The following are available online, Figure S1. ^{1}H-NMR spectrum of **1** (500 MHz in CDCl$_3$), Figure S2. ^{13}C-NMR spectrum of **1** (125 MHz in CDCl$_3$), Figure S3. DEPT-135°_spectrum of **1** (125 MHz in CDCl$_3$), Figure S4. HSQC Spectrum of **1**, Figure S5. ^{1}H-^{1}H COSY spectrum of **1**, Figure S6. HMBC spectrum of **1**, Figure S7. Infrared Spectrum of **1** (in KBr), Figure S8. HR-TOF-MS Spectrum of **1**, Figure S9. TLC Profile of **1**.

Author Contributions: Conceptualization, T.M. and U.S.; methodology, N.K.P. and Z.; software validation, S.F.; formal analysis, S.F.; investigation, R.M. and J.A.A.; resources, T.M. and M.Y.; data curation, N.K.P. and Z.; writing—original draft preparation, Z.; writing—review and editing, J.A.A. and M.Y.; visualization, R.M.; supervision, T.M., M.Y., and U.S.; project administration, N.K.P.; funding acquisition T.M. All authors have read and agreed to the published version of the manuscript.

Funding: This research was funded by PTM, Ministry of Research, Technology and Higher Education, Indonesia, grant number: 1827/UN6.3.1/LT/2020 (T.M.).

Acknowledgments: We thank to Joko Kusmoro, M.P. at Jatinangor Herbarium for identification of the plant material, Ahmad Darmawan at the Research Center for Chemistry, Indonesian Science Institute, for performing the NMR measurements, Kansi at the Center Laboratory of Universitas Padjadjaran for performing the HR-TOF-MS measurements, Tenny Putri Wikayani and Nurul Qomarilla at Cells and Tissues Culture Laboratory, Faculty of Medicine, Universitas Padjadjaran for the cytotoxic assay.

Conflicts of Interest: The authors declare no conflict of interest.

References

1. Techavuthiporn, C. *Langsat–Lansium Domesticum. Exotic Fruits Reference Guides*; Rodrigues, S., de Brito, E.S., Silva, E.d.O., Eds.; Academic Press: London, UK, 2018; pp. 279–283.
2. Mayanti, T.; Tjokronegoro, R.; Supratman, U.; Mukhtar, M.R.; Awang, K.; Hadi, A.H.A. Antifeedant triterpenoids from the seeds and bark of *Lansium domesticum* cv kokossan (Meliaceae). *Molecules* **2011**, *16*, 2785–2795. [CrossRef]
3. Dong, S.H.; Zhang, C.R.; Dong, L.; Wu, Y.; Yue, J.M. Onoceranoid-type triterpenoids from *Lansium domesticum*. *J. Nat. Prod.* **2011**, *74*, 1042–1048. [CrossRef] [PubMed]

4. Nishizawa, M.; Nishide, H.; Hayashi, Y.; Kosela, S. The structure of lansioside A: A novel triterpene glycoside with amino-sugar from *Lansium domesticum*. *Tetrahedron Lett.* **1982**, *23*, 1349–1350. [CrossRef]

5. Nishizawa, M.; Nishide, H.; Kosela, S.; Hayashi, Y. Structure of lansiosides: Biologically active new triterpene glycosides from *Lansium domesticum*. *J. Org. Chem.* **1983**, *48*, 4462–4466. [CrossRef]

6. Tanaka, T.; Ishibashi, M.; Fujimoto, H.; Okuyama, E.; Koyano, T.; Kowiyhayakorn, T.; Hayashi, M.; Komiyama, K. New onoceranoid constituents from *Lansium domesticum*. *J. Nat. Prod.* **2002**, *65*, 1709–1711. [CrossRef] [PubMed]

7. Habaguchi, K.; Watanabe, M.; Nakadaira, Y.; Nakanishi, K.; Kaing, A.K.; Lim, F.L. The full structures of lansic acid and its minor congener, an unsymmetric onoceradienedione. *Tetrahedron Lett.* **1986**, *34*, 3731–3734. [CrossRef]

8. Mayanti, T.; Supratman, U.; Mukhtar, M.R.; Awang, K.; Ng, S.W. Kokosanolide from the seed of *Lansium domesticum* Corr. *Acta Crystallogr.* **2009**, *E65*, o750.

9. Supratman, U.; Mayanti, T.; Awang, K.; Mukhtar, M.R.; Ng, S.W. 14-Hydroxy-8,14-secogammacera-7-ene-3,21-dione from the bark of *Lansium domesticum* Corr. *Acta Crystallogr.* **2010**, *E66*, o1621.

10. Manosroi, A.; Jantrawut, P.; Sainakham, M.; Manosroi, W.; Manosroi, J. Anticaner activities of the extract from longkong (*Lansium domesticum*) young fruits. *Pharm. Biol.* **2012**, *50*, 1397–1407. [CrossRef] [PubMed]

11. Ragasa, C.Y.; Labrador, P.; Rideout, J.A. Antimicrobial terpenoid from *Lansium domesticum*. *Philipp. Agric. Sci.* **2006**, *89*, 101–105.

12. Leatemia, J.A.; Isman, M.B. Insecticidal activity of crude seed extracts of *Annona.* spp., *Lansium domesticum* and *Sandoricum koetjape* against lepidopteran larvae. *Phytopatasitica* **2004**, *32*, 30–37. [CrossRef]

13. Saewan, N.; Sutherland, J.D.; Chantrapromma, K. Antimalarial tetranortriterpenoids from the seed of *Lansium domesticum* Corr. *Phytrochemistry* **2006**, *67*, 2288–2293. [CrossRef] [PubMed]

14. Matsumoto, T.; Kitagawa, T.; Teo, S.; Anai, Y.; Ikeda, R.; Imahori, D.; Ahmad, H.S.; Watanabe, T. Structures and antimutagenic effects of onoceranoid-type triterpenoids from the leaves of *Lansium domesticum*. *J. Nat. Prod.* **2018**, *81*, 2187–2194. [CrossRef] [PubMed]

15. Matsumoto, T.; Kitagawa, T.; Ohta, T.; Yoshida, T.; Imahori, D.; Teo, S.; Ahmad, H.S.; Watanabe, T. Structures of triterpenoids from the leaves of *Lansium domesticum*. *J. Nat. Med.* **2019**, *73*, 727–734. [CrossRef] [PubMed]

16. Mayanti, T.; Sianturi, J.; Harneti, D.; Darwati; Supratman, U.; Rosli, M.M.; Fun, H.K. 9,19-Cyclolanost-24-en-3-one,21,23-epoxy-21,22-dihydroxy (21*R*, 22*S*, 23*S*) from the leaves of *Lansium domesticum* Corr cv kokossan. *Molbank* **2015**, *2015*, M880. [CrossRef]

17. Hou, Y.; Cao, S.; Brodie, P.J.; Miller, J.S.; Birkinshaw, C.; Andrianjaty, M.N.; Andrlantsiferana, R.; Rasamison, V.E.; Tendyke, K.; Shen, Y.; et al. Euphane triterpenoids of *Cassipourea lanceolata* from the Madagascar rainforest. *Phytochemistry* **2010**, *71*, 669–674. [CrossRef] [PubMed]

Sample Availability: Samples of the compounds **1** are available from the authors.

Short Note

2-Propyl-*N'*-[1,7,7-trimethylbicyclo[2.2.1]hept-2-ylidene]pentanehydrazide

Mariia Nesterkina [1,*], Dmytro Barbalat [2], Ildar Rakipov [1,3] and Iryna Kravchenko [1]

[1] Department of Organic and Pharmaceutical Technologies, Odessa National Polytechnic University, 65044 Odessa, Ukraine; rakipovildar@gmail.com (I.R.); kravchenko.pharm@gmail.com (I.K.)

[2] Department of Analytical and Toxicological Chemistry, Odessa I.I. Mechnikov National University, 65082 Odessa, Ukraine; dmitriybar@ukr.net

[3] A.V. Bogatsky Physico-Chemical Institute, National Academy of Sciences of Ukraine, 65080 Odessa, Ukraine

* Correspondence: mashaneutron@gmail.com; Tel.: +38-093-713-3853

Received: 28 September 2020; Accepted: 19 October 2020; Published: 26 October 2020

Abstract: 2-Propyl-*N'*-[1,7,7-trimethylbicyclo[2.2.1]hept-2-ylidene]pentanehydrazide was obtained in 80% yield via the Einhorn variation of the Schotten–Baumann method by (+)-camphor hydrazide condensation with valproic acid (VPA) chloride. The structure of the titled compound was verified by Raman, FTIR, ^{1}H-NMR, and ^{13}C-NMR spectral analysis along with FAB-mass spectrometry. Thermal properties of synthesized derivative were elucidated by DSC and its purity by HPLC. The compound was successfully tested as a potential anticonvulsant agent based on models of chemically- and electrically-induced seizures.

Keywords: hydrazone; (+)-camphor; valproic acid; technology; terpenoid; anticonvulsant activity

1. Introduction

Valproic acid (VPA) refers to extensively used antiepileptic drugs (AEDs) that have been found to be effective against all types of seizures [1]. However, clinical use of VPA is associated with a significant adverse effect such as hepatotoxicity [2], causing the need for chemical modification of the VPA molecule aimed at dose reducing VPA-containing derivatives. In this context, particular interest is focused on terpenoids, which have proven to be anticonvulsant agents and serve as versatile scaffolds for enhancing the permeability of drugs [3,4]. Recently, this strategy has been applied to facilitate penetration of gamma-amino butyric acid (GABA) through the blood-brain barrier (BBB) by its conjugation with *l*-menthol, resulting in an increase in anticonvulsant potency [5]. Notably, the binding of biologically active molecules to terpenoids is expedient by the formation of enzymatically degradable bonds such as -NH-N=C-, -CO-OR-, -CO-NH-, etc.

Bearing the aforementioned in mind, bicyclic terpenoid (+)-camphor possessing its own antiseizure action [6] has been used as a base for the synthesis of conjugates containing the VPA moiety. Thus, we report herein on the synthesis and detailed structural determination of 2-propyl-*N'*-[1,7,7-trimethylbicyclo[2.2.1]hept-2-ylidene]pentanehydrazide comprising both (+)-camphor and VPA residues. The obtained compound was then investigated as a potential anticonvulsant agent on the models of maximal electroshock seizure (MES) and pentylenetetrazole (PTZ)-induced convulsions.

2. Results and Discussion

2.1. Chemistry

2-Propyl-*N'*-[1,7,7-trimethylbicyclo[2.2.1]hept-2-ylidene]pentanehydrazide **3** was synthesized via the Einhorn variation of the Schotten-Baumann method, as shown in Scheme 1. For this purpose, camphor hydrazine **1** was acylated with valproic acid chloride **2** in the presence of triethylamine (TEA) both as an acid scavenger and nucleophilic acylation catalyst. Initial hydrazine **1** was obtained by camphor treatment with hydrazine hydrate and acid chloride **2** through VPA reflux with $SOCl_2$ [7,8]. Target compound was isolated in an 80% yield as white crystals well soluble in organic solvents such as chloroform, hexane, acetonitrile, and benzene. The structure of hydrazone **3** was verified by ^{13}C-NMR, ^{1}H-NMR, FT-IR, Raman spectroscopy, and FAB-mass spectrometry. Thermal transition of compound **3** was carried out by differential scanning calorimetry (DSC) along with the determination of melting enthalpy (ΔH_m) as 120.2 J/g. The purity was assessed by HPLC analysis using the reversed-phase method with isocratic elution by a system composed of acetonitrile: 0.01% formic acid aqueous solution (70:30). According to the HPLC assay, hydrazone purity was established by internal normalization with detection at ultraviolet wavelengths (230 nm and 260 nm) as 100% and 99%, respectively.

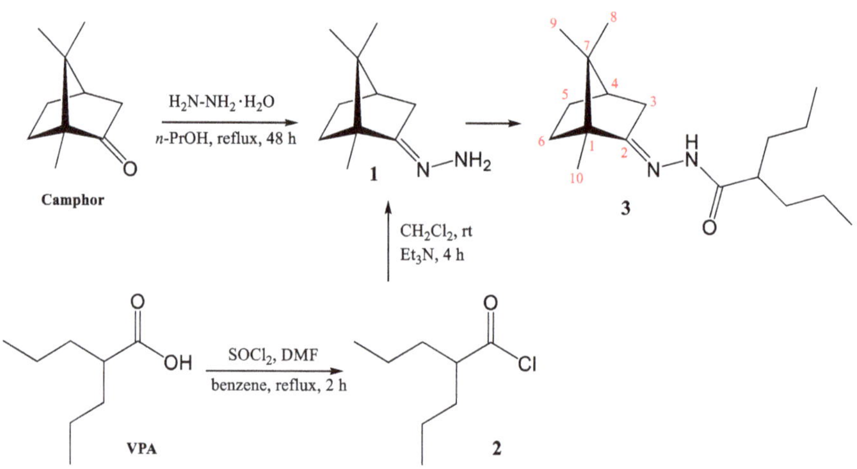

Scheme 1. Synthesis of 2-propyl-*N*′-[1,7,7-trimethylbicyclo[2.2.1]hept-2-ylidene]pentanehydrazide.

Fast-atom bombardment (FAB) has been applied as an ionization method in mass spectrometry investigation; the FAB-MS spectrum of derivative **3** displays the protonated molecular ion peak [M + H]$^+$ at *m/z* 293. The FT-IR spectrum exhibits absorption bands of N–H bonds (3179 cm^{-1}), C=O groups (1697 cm^{-1}), and alkyl C–H (2970-2834 cm^{-1}). Taking into account the overlapping of C=N and C=O vibrations in the FT-IR spectrum, the formation of the imine (C=N) group was confirmed by Raman spectroscopy as an intense peak at 1646 cm^{-1}. The ^{1}H-NMR spectrum corroborates the proposed structure of compound **3** by chemical shift, integration, and multiplicity of resonance signals. The characteristic proton signal of the -NH group was observed as a singlet at 9.51 ppm. Likewise, the ^{13}C-NMR spectrum detected the presence of all carbon atoms in the molecule.

2.2. Anticonvulsant Activity

Anticonvulsant activity of the title compound was determined by two pharmacological seizure models including intravenous pentylenetetrazole (i.v. PTZ) and maximal electroshock seizure (MES) tests. In the PTZ-induced model, the anticonvulsant effect was established by the assessment of PTZ minimum effective doses (MED) that provoke clonic-tonic convulsions (DCTC) and tonic extension (DTE). As illustrated in Figure 1, camphor, hydrazone **3**, along with VPA, demonstrated protection against PTZ-induced seizures at 3 h after administration, as validated by the increase of DCTC and DTE values to 165–187% and 170–197%, respectively, compared with the control (100%). However, at this time point, there were no statistically significant differences between the afore-mentioned experimental groups, indicating comparable activity of the investigated compounds. In marked contrast, synthesized derivative **3** was found to possess significant antiseizure action over a long time period (24 h after administration) with the average values of 277% for DCTC and 238% for DTE. As seen, hydrazone **3** exhibited higher potency versus initial camphor and VPA ($p < 0.01$), which highlighted its prolonged effect.

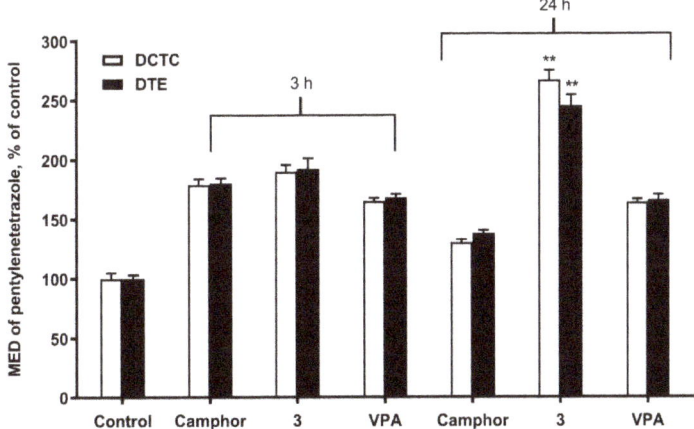

Figure 1. At 3 h and 24 h after oral administration. Values are given as mean ± SEM, $n = 5$ mice; for all groups $p < 0.01$ compared with the control; ** $p < 0.01$ compared with VPA.

In the MES test, camphor hydrazone **3** substantially prevented the animals' mortality at 3 h after administration, manifesting 80% protection that is equivalent to the VPA effect (80%), whereas moderate antiseizure action was observed for the initial camphor (60%) (Table 1). Conversely, the activity of the titled compound was maintained at a long time period (24 h) with 100% of mortality protection, which confirms the idea toward enzymatic cleavage of labile bonds (C=N, N-NH, or CO-NH) in hydrazone molecules, followed by gradual release of pure terpenoid and VPA.

Table 1. Against maximal electroshock (MES)-induced seizures.

Compound	Control	Camphor	VPA	Compound 3
3 h after single oral administration				
% Mortality protection	0	60	80	80
24 h after single oral administration				
% Mortality protection	0	20	60	100

Thus, camphor derivative **3** protects against seizures induced by chemical and electrical stimuli both in short (3 h) and long (24 h) time periods.

3. Materials and Methods

3.1. General Information

(+)-Camphor and valproic acid (VPA) were obtained from commercial sources. Hydrazine **1** and acid chloride **2** were synthesized according to the standard procedure [7,8]. The progress of reaction was monitored by TLC on Merck-made (TLC Silica gel 60 F_{254}) plates (Darmstadt, Germany) visualized by UV light using ethyl acetate-benzene (1:1) as the eluent system. Structure of the obtained compound was established by ^{1}H-NMR spectroscopy on a Varian VXR-300 (300 MHz) instrument (Varian Inc., Palo Alto, CA, USA) and by ^{13}C-NMR spectroscopy on a Varian-Mercury 400 spectrometer

(Varian Inc., Palo Alto, CA, USA) using DMSO-d_6 as a solvent and TMS as an internal standard. FAB mass spectra were obtained on a VG 70-70EQ mass spectrometer (VG Analytical Ltd., Manchester, UK) equipped with a Xe ion gun (8 kV); the sample was mixed with the *m*-nitrobenzyl-alcohol matrix. The purity of the compound was checked by high-performance liquid chromatography on an Agilent 1260 Infinity HPLC system (Agilent, Santa Clara, CA, USA). IR spectra were measured with a Frontier FT-IR spectrometer (Perkin-Elmer, Hopkinton, MA, USA) using KBr pellets. Raman spectra were undertaken with a DXR Raman Microscope (Thermo Fisher Scientific, Madison, WI, USA). DSC curves were recorded in a Q2000 differential scanning calorimeter (TA Instruments, New Castle, DE, USA) using aluminum crucibles containing approximately 2 mg of samples, under a dynamic nitrogen atmosphere and a heating rate of 5 °C min^{-1} in the temperature range of 20 to 200 °C.

3.2. Synthesis of 2-Propyl-N′-[1,7,7-trimethylbicyclo[2.2.1]hept-2-ylidene]pentanehydrazide (3)

To a stirred solution of (+)-camphor hydrazine **1** (0.8 g, 5.26 mmol) in CH_2Cl_2 (25 mL) at room temperature, valproic acid chloride (0.897 g, 5.52 mmol) was added. The reaction mixture was cooled to 0 °C, stirred for 10 min, and triethylamine (TEA) was added dropwise (0.77 mL, 5.52 mmol). Stirring was continued for 30 min, then the flask was gradually warmed to room temperature and the stirring continued for an additional 4 h. The reaction completion was monitored by TLC. The reaction mixture was filtered, the filtrate was diluted with CH_2Cl_2 to 100 mL, and washed with 1 M aqueous HCl, 10% aqueous $NaHCO_3$, and water. The combined organic phases were dried over anhydrous Na_2SO_4 and concentrated under reduced pressure. The crude product was purified by recrystallization from methanol.

White crystals (80%). ^{1}H-NMR (300 MHz, DMSO-d_6) δ: 0.65 (s, 3H, CH_3), 0.82 (s, 6H, 2CH_3), 0.86 (s, 3H, CH_3), 0.90 (s, 3H, CH_3), 1.21 (m, 8H, 2CH_2-CH_2), 1.46-1.51 (m, 2H, CH_2), 1.65 (t, 1H, CH), 1.77 (m, 1H, CH), 1.89 (m, 2H, CH_2), 2.30 (m, 1H, CH), 9.51 (s, 1H, NH). ^{13}C-NMR (100 MHz, DMSO-d_6) δ: 177.4 (C=O), 171.3 (C-2), 52.3 (C-1), 47.8 (C-7), 47.6 (C-4), 43.9 (CH), 35.1 (C-3), 34.5 (CH_2), 32.7 (C-6), 27.3 (C-5), 20.5 (CH_2), 18.8 (C-8,9), 14.2 (C-10), 11.5 (CH_3). FT-IR ($v_{max, CM}$$^{-1}$): 3179 (N-H), 2970–2834 (C-H), 1697 (C=O). Raman ($v_{max, CM}$$^{-1}$): 2943-2872 (C-H), 1646 (C=N). MS (FAB) *m/z*: 293 [M + H]$^+$. HPLC: t_r = 23.43 min. M.p. (DSC) onset: 154.26 °C, peak max: 155.46 °C (Supplementary Materials).

3.3. Anticonvulsant Screening

Anticonvulsant activity was studied using outbreed male white mice (18–22 g) as experimental animals. Animals were maintained under a 12 h light regime and in a standard animal facility with free access to water and food. All the animals were purchased from Odessa National Medical University, Ukraine. The Animal Ethics Committee (agreement No. 03/2020) of Odessa National Polytechnic University (Ukraine) approved the study. (+)-Camphor, VPA, and compound **3** were administered orally (preliminarily dissolved in in Tween 80/water emulsion): camphor at a dose of 50 mg/kg; VPA and hydrazone **3** in equimolar amounts. Antiseizure evaluation was carried out at 3 h and 24 h after administration both on MES and PTZ-induced convulsions according to the earlier reported procedures [9,10].

4. Conclusions

Einhorn variation of the Schotten–Baumann method was successfully applied for the synthesis of 2-propyl-N′-[1,7,7-trimethylbicyclo[2.2.1]hept-2-ylidene]pentanehydrazide via (+)-camphor hydrazide condensation with valproic acid (VPA) chloride, followed by structure confirmation using Raman, FT-IR, ^{1}H-NMR, and ^{13}C-NMR spectral analysis along with FAB-mass spectrometry. The title compound was found to demonstrate prolonged anticonvulsant activity both on PTZ- and MES-induced seizures.

Supplementary Materials: Copies of the ^{1}H-NMR ^{13}C-NMR, FT-IR, Raman, FAB mass spectra, DSC thermograms, and HPLC chromatograms are available online.

Molbank **2020**, *2020*, M1164

Author Contributions: I.K. conceived and designed the experiments; M.N. performed the synthesis and analyzed the NMR spectral data; D.B. performed the analysis of FT-IR, Raman, DSC, and HPLC experiments; I.R. carried out FAB characterization of the compound. All authors contributed in manuscript writing. All authors have read and agreed to the published version of the manuscript.

Funding: This research received no external funding.

Conflicts of Interest: The authors declare no conflict of interest.

References

1. Romoli, M.; Mazzocchetti, P.; D'Alonzo, R.; Siliquini, S.; Rinaldi, V.E.; Verrotti, A.; Calabresi, P.; Costa, C. Valproic acid and epilepsy: From molecular mechanisms to clinical evidences. *Curr. Neuropharmacol.* **2019**, *17*, 926–946. [CrossRef]

2. Gayam, V.; Mandal, A.K.; Khalid, M.; Shrestha, B.; Garlapati, P.; Khalid, M. Valproic acid induced acute liver injury resulting in hepatic encephalopathy- a case report and literature review. *J. Community Hosp. Intern. Med. Perspect.* **2018**, *8*, 311–314. [CrossRef] [PubMed]

3. De Almeida, R.N.; Agra, M.; Maior, F.N.; De Sousa, D.P. Essential oils and their constituents: Anticonvulsant activity. *Molecules* **2011**, *16*, 2726–2742. [CrossRef] [PubMed]

4. Chen, J.; Jiang, Q.D.; Chai, Y.P.; Zhang, H.; Peng, P.; Yang, X.X. Natural terpenes as penetration enhancers for transdermal drug delivery. *Molecules* **2016**, *21*, 1709. [CrossRef] [PubMed]

5. Nesterkina, M.V.; Kravchenko, I.A. Synthesis and anticonvulsant activity of menthyl γ-aminobutyrate. *Chem. Nat. Compd.* **2016**, *52*, 237–239. [CrossRef]

6. Agrawal, S.; Jain, J.; Kumar, A.; Gupta, P.; Garg, V. Synthesis molecular modeling and anticonvulsant activity of some hydrazone, semicarbazone, and thiosemicarbazone derivatives of benzylidene camphor. *Res. Rep. Med. Chem.* **2014**, *4*, 47–58. [CrossRef]

7. Da Silva, E.T.; da Silva Araújo, A.; Moraes, A.M.; de Souza, L.A.; Silva Lourenço, M.C.; de Souza, M.V.; Wardell, J.L.; Wardell, S.M. Synthesis and biological activities of camphor hydrazone and imine derivatives. *Sci. Pharm.* **2016**, *84*, 467–483. [CrossRef] [PubMed]

8. Wang, Z.; Li, J.; Zeng, X.D.; Hu, X.M.; Zhou, X.; Hong, X. Synthesis and pharmacological evaluation of novel benzenesulfonamide derivatives as potential anticonvulsant agents. *Molecules* **2015**, *20*, 17585–17600. [CrossRef] [PubMed]

9. Nesterkina, M.V.; Alekseeva, E.A.; Kravchenko, I.A. Synthesis, physicochemical properties, and anticonvulsant activity of the gaba complex with a calix[4]arene derivative. *Pharm. Chem. J.* **2014**, *48*, 82–84. [CrossRef]

10. Nesterkina, M.; Barbalat, D.; Konovalova, I.; Shishkina, S.; Atakay, M.; Salih, B.; Kravchenko, I. Novel (–)-carvone derivatives as potential anticonvulsant and analgesic agents. *Nat. Prod. Res* **2020**, (in press). [CrossRef] [PubMed]

MDPI

Short Note

N-[7-Chloro-4-[4-(phenoxymethyl)-1H-1,2,3-triazol-1-yl]quinoline]-acetamide

Paolo Coghi [1,2,*], Jerome P. L. Ng [2,3], Ali Adnan Nasim [2,3] and Vincent Kam Wai Wong [2,3,*]

[1] School of Pharmacy, Macau University of Science and Technology, Macau 999078, China
[2] State Key Laboratory of Quality Research in Chinese Medicine, Macau University of Science and Technology, Macau 999078, China; plng@must.edu.mo (J.P.L.N.); aanasim@gmail.com (A.A.N.)
[3] Dr. Neher's Biophysics Laboratory for Innovative Drug Discovery, State Key Laboratory of Quality Research in Chinese Medicine, Macau University of Science and Technology, Macau 999078, China
* Correspondence: coghips@must.edu.mo (P.C.); bowaiwong@gmail.com (V.K.W.W.); Tel.: +86-(853)-8897-2408 (V.K.W.W.)

Abstract: The 1,2,3-triazole is a well-known biologically active pharmacophore constructed by the copper-catalyzed azide–alkyne cycloaddition. We herein reported the synthesis of 4-amino-7-chloro-based [1,2,3]-triazole hybrids via Cu(I)-catalyzed Huisgen 1,3-dipolar cycloaddition of 4-azido-7-chloroquinoline with an alkyne derivative of acetaminophen. The compound was fully characterized by Fourier-transform infrared (FTIR), proton nuclear magnetic resonance (^{1}H-NMR), carbon-13 nuclear magnetic resonance (^{13}C-NMR), heteronuclear single quantum coherence (HSQC), ultraviolet (UV) and high-resolution mass spectroscopies (HRMS). This compound was screened in vitro with different normal and cancer cell lines. The drug likeness of the compound was also investigated by predicting its pharmacokinetic properties.

Keywords: 1,2,3-triazole; anticancer; aminoquinoline; hybrid compound

Citation: Coghi, P.; Ng, J.P.L.; Nasim, A.A.; Wong, V.K.W. N-[7-Chloro-4-[4-(phenoxymethyl)-1H-1,2,3-triazol-1-yl]quinoline]-acetamide. *Molbank* **2021**, *2021*, M1213. https://doi.org/10.3390/M1213

Academic Editor: Bartolo Gabriele

Received: 20 April 2021
Accepted: 13 May 2021
Published: 20 May 2021

Publisher's Note: MDPI stays neutral with regard to jurisdictional claims in published maps and institutional affiliations.

1. Introduction

The 7-chloroquinoline moiety, a pharmacophore of several established antimalarial drugs such as chloroquine (Figure 1a) [1], is recently being focused on as a potential anti-cancer agent as well as a chemosensitizer when used in combination with anti-cancer drugs [2].

With increasing drug resistance to available agents, intensive drug discovery efforts on developing new antimalarial/anticancer drugs or modifying existing agents are ongoing [3]. Molecular hybridization as a drug discovery strategy involves the rational design of new chemical entities by the fusion (usually via a covalent linker) of two drugs, of which both active compounds and/or pharmacophoric units are recognized and derived from known bioactive molecules [3].

In order to broaden the structural diversity of the compounds and intensify their biological activities, covalently linked hybrids were created with a pharmacologically significant class of compounds known as [1,2,3]-triazole (Figure 1b–d). These [1,2,3]-triazoles have exhibited a myriad of biological activities including antifungal, [4] antibacterial [5] and antitubercular activities [6]. Some structures of quinolone- and triazole-based hybrids reported in literature [7–10] (Figure 1e–h).

We have previously reported the conversion of the commercially available 4,7-dichloroquinoline **1** to a series of 4-amino-7-chloroquinolone derivatives [11]. Herein, we reported the synthesis of a novel 4-amino-7-chloroquinoline-based 1,2,3-triazole hybrid **5** by Cu(I)-catalyzed azide–alkyne cycloaddition. The structure of compound **5** was characterized by NMR, MS, FT-IR and UV spectra. The cytotoxicity of **5** was also evaluated against different cell lines.

27

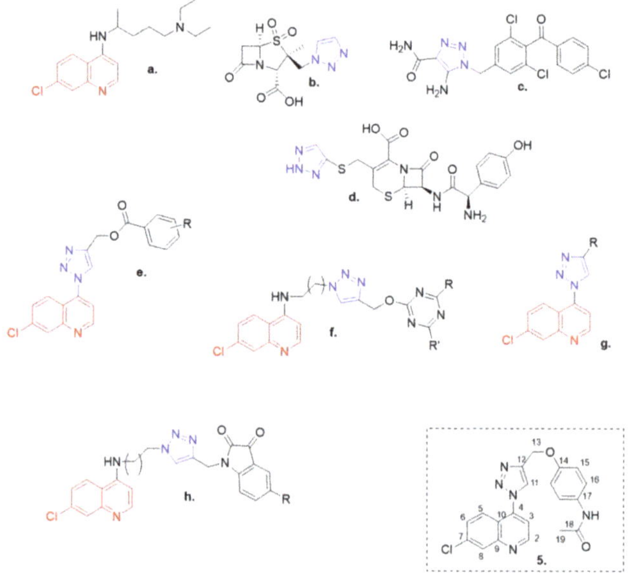

Figure 1. Clinically approved quinoline (red) and 1,2,3-triazole (blue) containing drugs. (**a**) Chloroquine; (**b**) tazobactam; (**c**) carbonic anhydrase inhibitors (CAI); (**d**) cefatrizine; (**e**–**h**) some structures of quinolone- and triazole-based hybrids reported in literature and compound **5** reported in this paper.

2. Results and Discussion

The synthesis of the triazole hybrid compound, *N*-[7-chloro-4-[4-(phenoxymethyl)-1*H*-1,2,3-triazol-1-yl] quinoline]-acetamide **5**, involved the initial synthesis of the precursors 4-azido-7-chloroquinoline **2** and *O*-acetylenic acetaminophen **4**.

Quinoline **2** was furnished by applying the modified method reported by de Souza et al. [12], whereby 4,7-dichloroquinoline **1** reacted with two equivalents of NaN$_3$ in anhydrous DMF at 65 °C for 6 h (Scheme 1a). The recrystallization of the crude product from CH$_2$Cl$_2$/hexane afforded **2** in an 86% yield.

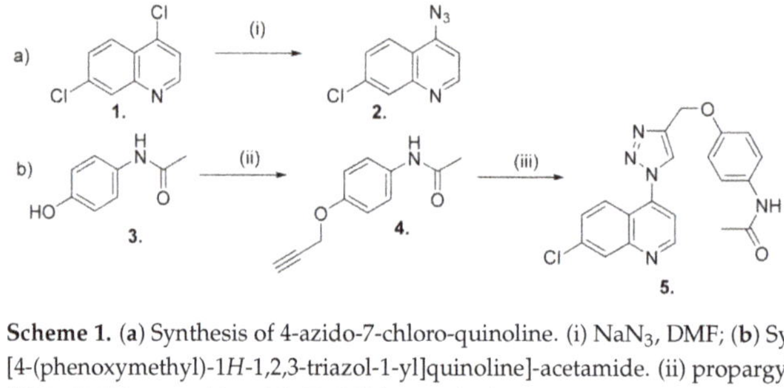

Scheme 1. (**a**) Synthesis of 4-azido-7-chloro-quinoline. (i) NaN$_3$, DMF; (**b**) Synthesis of *N*-[7-chloro-4-[4-(phenoxymethyl)-1*H*-1,2,3-triazol-1-yl]quinoline]-acetamide. (ii) propargyl bromide, DMF, K$_2$CO$_3$; (iii) **2**, CuSO$_4$, ascorbic acid, *t*BuOH/water (1:1).

The *O*-alkylation reaction of acetaminophen **3** with 1.5 equivalents of propargyl bromide and anhydrous K$_2$CO$_3$ in anhydrous DMF yielded the acetylenic intermediate **4** in a good yield after the recrystallization from CH$_2$Cl$_2$/hexane (82%, Scheme 1b).

The hybrid compound **5** was finally obtained by using a modified cycloaddition procedure reported by Fokin et al. [13].

Equimolars of quinoline **2** and acetylene **4** were dissolved in *t*BuOH/water (1:1) and were treated with sodium ascorbate (0.4 equiv.) and CuSO$_4$ (20 mol%) sequentially

(Scheme 1b). The reaction mixture was then stirred vigorously at 65 °C for 24 h. Compound 5 was isolated by column chromatography in a high yield (72%).

The structure of **5** was verified by ^{1}H and ^{13}C NMR spectra (Supplementary Materials, Figures S3 and S4). The ^{1}H-NMR spectrum showed a singlet at 5.33 ppm associated with the C13-methylene group and a singlet at 2.07 ppm associated with the α-position of the C18-carbonyl group.

Regarding the ^{13}C NMR signals, the disappearance of characteristic peaks of acetylenic group at 78.5 and 75.5 ppm, and the appearance of a C-13 methylene signal at 61.5 ppm and C-11 vinylic signal at 127.3 ppm were probably the most relevant features to verify the incorporation of a triazole moiety.

Heteronuclear single quantum coherence spectroscopy (HSQC) was used to assign ^{13}C signals of compound **5** as shown in Table S1 (see Supplementary Materials for 2D spectra, Figures S5 and S6).

The IR spectrum of compound **5** showed characteristic N–H stretching at 3448 cm^{-1}, C=O stretching at 1674 cm^{-1} and C=C stretching at 1612 cm^{-1} (Supplementary Materials, Figure S7). Other vibrational peaks at 1550, 1512 and 1411 cm^{-1} were corresponding to the N–H bending in the amide. Moreover, the absence of the azide stretching peak of **2** at 2123 cm^{-1} in the IR spectrum of **5** confirmed the conversion of azide.

The UV and HRMS of **5** was also recorded for further characterization (Supplementary Materials, Figures S8 and S9).

The target compound **5** showed a low cytotoxicity by evaluating its in vitro cytotoxic activities against normal (LO$_2$ and BEAS-2B), immortalized (HEK293) and cancer (HepG2 and A549) cell lines (IC$_{50}$ values > 100 μM, Supplementary Materials Figure S10).

In addition, the promising compound **5** was assessed by predicting its physicochemical properties and oral bioavailability. From the calculated physicochemical properties (Supplementary Materials Table S2), compound **5** did not violate any of Lipinski's rules [14], indicating its drug-like character and a good chance for oral administration.

This finding corroborates the results of the gastrointestinal absorption from SwissADME, in which compound 5 was predicted with high absorption according to BOILED-Egg model [15] (Figure S11) and data from admetSAR 2 [16], in which the compound was predicted to be orally bioavailable and absorbed in human intestine.

3. Materials and Methods

3.1. Chemistry

Silica gel (FCP 230–400 mesh) was used for column chromatography. Thin-layer chromatography was carried out on E. Merck precoated silica gel 60 F$_{254}$ plates and visualized with phosphomolybdic acid, iodine, or a UV-visible lamp.

All chemicals were purchased from Bide Pharmatech., Ltd. (Shanghai, China) and J & K scientific (Hong Kong, China). ^{1}H-NMR and ^{13}C-NMR spectra were collected in CDCl$_3$ at 25 °C on a Bruker Ascend®-600 NMR spectrometer (600 MHz for ^{1}H and 150 MHz for ^{13}C) (Bruker, Billerica, MA, USA). All chemical shifts were reported in the standard δ notation of parts per million using the peak of residual proton signals of CDCl$_3$ or DMSO-d6 as an internal reference (CDCl$_3$, δ_C 77.2 ppm, δ_H 7.26 ppm; DMSO-d6, δ_C 39.5 ppm, δ_H 2.50 ppm). High-resolution mass spectra (HRMS) were measured using electrospray ionization (ESI). The measurements were done in a positive ion mode (interface capillary voltage 4500 V); the mass ratio was from m/z 50 to 3000 Da; external/internal calibration was done with electrospray calibration solution.

HRMS analyses were performed by an Agilent 6230 electrospray ionization (ESI) time-of-flight (TOF) mass spectrometer with Agilent C18 column (4.6 mm × 150 mm, 3.5 μm). The mobile phase was isocratic (water +0.01% TFA; CH$_3$CN) at a flow rate of 0.5 mL/min. The peaks were determined at 254 nm under UV.

UV analysis was performed by a Shimadzu UV—2600 (Osaka, Japan) with 1 cm quartz cell and a slit width of 2.0 nm. The analysis was carried out using a wavelength in the range of 200–400 nm.

IR analysis (KBr) was performed by a Shimadzu IRAffinity-1S (Osaka, Japan) with a frequency range of 4000–500 cm^{-1}.

3.1.1. Synthesis of 4-Azido-7-chloro-quinoline (2)

The 4,7-Dichloroquinoline (2.0 g, 10 mmol) was dissolved in 5 mL anhydrous DMF. NaN$_3$ (1.3 g, 20 mmol) was then added in one portion, and the resulting mixture was stirred at 65 °C for 6 h, whereupon TLC indicated reaction completion. The reaction mixture was then allowed to cool to ambient temperature, after which it was diluted with 100 mL CH$_2$Cl$_2$, washed with water (3 × 30 mL), dried over anhydrous Na$_2$SO$_4$, and evaporated to dryness. The resulting product residue was recrystallized from a CH$_2$Cl$_2$/hexane 1:1 mixture to yield the final pure product **2** as colorless, needle-like crystals in 86% yield. δ_H (600 MHz, CDCl$_3$) 8.82 (1H, d, *J* = 4.9 Hz, H-2), 8.09 (1H, d, *J* = 2.4 Hz, H-8), 8.01 (1H, d, *J* 9.3, H-5), 7.49 (1H, dd, *J* 2.4 and 9.3, H-6) 7.12 (1H, d, *J* 4.9, H-3) ppm; δ_C (150 MHz, CDCl$_3$) 150.9, 149.1, 146.8, 136.9, 127.9, 123.8, 119.9, 108.7 ppm. The spectral characteristics are consistent with those of **2** in the literature [17].

3.1.2. Synthesis of N-[4-(propargyloxy) Phenyl] acetamide (4)

Acetaminophen (*N*-acetyl-*para*-aminophenol) (13 mmol) was dissolved in 10 mL of anhydrous DMF. Anhydrous K$_2$CO$_3$ (2.7 g, 19.5 mmol) was then added to the solution, and the mixture was stirred at 30 °C for 30 min. Propargyl bromide (3-bromopropyne, 2.2 mL, 19.5 mmol) was then added slowly to the reaction mixture, and subsequently stirred at 30 °C for 6 h upon which TLC indicated completion of the reaction. The reaction mixture was then diluted with 50 mL water and extracted with ethyl acetate (3 × 50 mL). These extracts were then combined, washed with water (2 × 50 mL), dried over anhydrous Na$_2$SO$_4$ and evaporated in vacuo to yield the product residue that was then recrystallized from the CH$_2$Cl$_2$/hexane 1:1 mixture to yield the compound 4 in 82% yield. δ_H (600 MHz, CDCl$_3$) 2.15 (3H, s, Me), 2.51 (1H, s), 4.67 (2H, s, CH$_2$), 6.94 (2H, d, *J* = 8.99 Hz, H$_{Ar}$), 7.22 (1H, br, NH), 7.40 (2H, d, *J* = 8.99 Hz, H$_{Ar}$) ppm; δ_C (150MHz, CDCl$_3$) 24.2, 56.3, 75.5, 78.5, 115.5, 121.8, 131.9, 154.6, 168.3 ppm. The spectral characteristics are consistent with those of **4** in the literature [18].

3.1.3. Synthesis of N-[7-chloro-4-[4-(phenoxymethyl)-1H-1,2,3-triazol-1-yl] quinoline]-acetamide (5)

The derivative of acetaminophen 4 (1mmol) and the appropriate azide **2** were dissolved in 5 mL *t*BuOH/water (1:1) and, while stirring at 65 °C, 1 M sodium ascorbate (0.4 mL, 0.4 mmol) and 1 M CuSO$_4$ (0.2 mL, 20 mol%) were added sequentially, in that order. The reaction mixture was then stirred at 65 °C for 24 h. The crude product was then precipitated out by slowly adding cold water to the reaction mixture, after which it was filtered, washed with water, air dried and purified by silica column chromatography (eluents ranging in polarity from EtOAc/hex 3:7 to 5% MeOH in EtOAc). Yield 72%, δ_H (600 MHz, CDCl$_3$) 2.07 (3H, s, Me), 5.33 (2H, s, CH$_2$-O), 7.09 (2H, d, *J* = 9 Hz, H-15), 7.57 (2H, d, *J* = 9 Hz, H-16), 7.86 (1H, dd, *J* = 9.1 and 2.2 Hz, H-6), 7.93 (1H, d, *J* = 4.65 Hz, H-3), 8.03 (1H, d, *J* = 9.1 Hz, H-5), 8.36 (1H, d, *J* = 2.3 H-8), 9.01 (1H, s, H-11), 9.23 (1H, d, *J* = 4.65 Hz, H-2), 9.89 (1H, s, NH) ppm; δ_C (150MHz, CDCl$_3$) 24.2 (C-19), 61.7 (C-13), 115.3 (C-15), 117.6 (C-3), 120.8 (C-16), 121 (C-10), 125.8 (C-5), 127.3 (C-11), 128.6 (C-8), 129.5 (C-6), 133.5 (C-17), 135.9 (C-4), 140.8 (C-7), 144.2 (C-12), 149.8 (C-9), 152.8 (C-2), 154.1 (C-14), 168.2 (CO) ppm; HRMS-ESI *m/z* 394.1083 [M + H]$^+$ (calculated for C$_{20}$H$_{17}$ClN$_2$O$_2$, *m/z* 394.1065); UV (CH$_2$Cl$_2$) peaks 211, 234 and 287 nm; IR (KBr) (ν_{max}/cm^{-1}) 3478 (NH), 2924 (CH), 2368, 1674 (NH amide), 1612, 1550, 1512, 1458, 1411, 1373, 1319, 1242, 1049, 825 cm^{-1}.

3.2. Biological Studies

Compound **5** was dissolved in DMSO at a final concentration of 50 mM and stored at −20 °C before use. Cytotoxicity was assessed by using the 3-(4,5-dimethylthiazole-2yl)-2,5-diphenyltetrazolium bromide (MTT) (5 mg/mL) assay as previously described [19]. Briefly, 4 × 10^3 cells per well were seeded in 96-well plates before drug treatments. After

Molbank **2021**, *2021*, M1213

overnight cell culture, the cells were then exposed to different concentrations of selected compounds (0.19–100 μM) for 72 h. Cells without drug treatment were used as controls. Subsequently, 10 μL of 5 mg/mL MTT solution was added to each well and incubated at 37 °C for 4 h followed by addition of 100 μL solubilization buffer (10 mM HCl in solution of 10% of SDS) and overnight incubation. Then A_{570} nm was determined in each well on the next day. The percentage of cell viability was calculated using the following formula: cell viability (%) = $A_{treated} / A_{control} \times 100$. A representative graph of at least three independent experiments was shown in Supplementary Materials Figure S8.

4. Conclusions

The synthesis of a potential triazole-based quinoline was presented. The chemical structure of the synthesized compound was verified by using NMR, mass, IR and UV spectrometries. The cytotoxicity and drug likeness of the compound were also determined by MTT assay and computations, respectively.

Supplementary Materials: The following are available online, Figure S1: ^{1}H NMR compound **4**, Figure S2: ^{13}C NMR compound **4**, Figure S3: ^{1}H NMR compound **5**, Figure S4: ^{13}C NMR compound **5**, Figures S5 and S6: HSQC compound **5**, Figure S7: IR spectrum compound **5**, Figure S8: HRMS of **5**, Figure S9: UV spectrum, Figure S10: cytotoxicity results, Table S1: ^{1}H and ^{13}C-nuclear magnetic spectroscopy (NMR) chemical shifts, Table S2: physicochemical properties of **5** calculated by SwissADME, Figure S11: BOILED-Egg graph.

Author Contributions: Conceptualization, P.C.; methodology, P.C. and J.P.L.N.; investigation, A.A.N.; data curation, J.P.L.N.; writing—original draft preparation, P.C.; writing—review and editing, J.P.L.N. and V.K.W.W.; supervision, P.C. and V.K.W.W.; project administration, P.C.; funding acquisition, P.C. All authors have read and agreed to the published version of the manuscript.

Funding: This research was funded by FDCT grants from Macao Science and Technology Development Fund to PC (project code: 0096/2020/A, 0087/2020/A).

Institutional Review Board Statement: Not applicable.

Informed Consent Statement: Not applicable.

Data Availability Statement: Not applicable.

Acknowledgments: The authors are grateful to Giovanni Ribaudo for his kind and grateful support and Yuhan Xie for support during analysis.

Conflicts of Interest: The authors declare no conflict of interest.

References

1. Pinheiro, L.C.S.; Feitosa, L.M.; Gandi, M.O.; Silveira, F.F.; Boechat, N. The Development of Novel Compounds Against Malaria: Quinolines, Triazolpyridines, Pyrazolopyridines and Pyrazolopyrimidines. *Molecules* **2019**, *24*, 4095. [CrossRef] [PubMed]
2. Sasaki, K.; Tsuno, N.H.; Sunami, E.; Tsurita, G.; Kawai, K.; Okaji, Y.; Nishikawa, T.; Shuno, Y.; Hongo, K.; Hiyoshi, M.; et al. Chloroquine potentiates the anti-cancer effect of 5-fluorouracil on colon cancer cells. *BMC Cancer* **2010**, *10*, 1–11. [CrossRef] [PubMed]
3. Viegas, C.; Danuello, A.; Bolzani, V.S.; Barreiro, E.J.; Fraga, C.A.M. Molecular hybridization: A useful tool in the design of new drug prototypes. *Curr. Med. Chem.* **2007**, *14*, 1829–1852.
4. Aher, N.G.; Pore, V.S.; Mishra, N.N.; Kumar, A.; Shukla, P.K.; Sharma, A.; Bhat, M.K. Synthesis and antifungal activity of 1,2,3-triazole containing fluconazole analogues. *Bioorg. Med. Chem. Lett.* **2009**, *19*, 759–763. [CrossRef] [PubMed]
5. Demaray, J.A.; Thuener, J.E.; Dawson, M.N.; Sucheck, S.J. Synthesis of triazole-oxazolidinones via a one-pot reaction and evaluation of their antimicrobial activity. *Bioorg. Med. Chem. Lett.* **2008**, *18*, 4868–4871. [CrossRef] [PubMed]
6. Tripathi, R.P.; Yadav, A.K.; Ajay, A.; Bisht, S.S.; Chaturvedi, V.; Sinha, S.K. Application of Huisgen (3 + 2) cycloaddition reaction: Synthesis of 1-(2,3-dihydrobenzofuran-2-yl-methyl [1,2,3]-triazoles and their antitubercular evaluations. *Eur. J. Med. Chem.* **2010**, *45*, 142–148. [CrossRef] [PubMed]
7. Pereira, G.R.; Brandão, G.C.; Arantes, L.M.; de Oliveira, H.A., Jr.; de Paula, R.C.; do Nascimento, M.F.; dos Santos, F.M.; da Rocha, R.K.; Lopes, J.C.; de Oliveira, A.B. 7-Chloroquinolinotriazoles: Synthesis by the azide-alkyne cycloaddition click chemistry, antimalarial activity, cytotoxicity and SAR studies. *Eur. J. Med. Chem.* **2014**, *73*, 295–309. [CrossRef] [PubMed]
8. Manohar, S.; Khan, S.I.; Rawat, D.S. Synthesis of 4-aminoquinoline-1,2,3-triazole and 4-aminoquinoline-1,2,3-triazole-1,3,5-triazine hybrids as potential antimalarial agents. *Chem. Biol. Drug. Des.* **2011**, *78*, 124–136. [CrossRef] [PubMed]

9. Boechat, N.; Ferreira, M.d.L.; Pinheiro, L.C.; Jesus, A.M.; Leite, M.M.; Júnior, C.C.; Aguiar, A.C.; de Andrade, I.M.; Krettli, A.U. New compounds hybrids 1h-1,2,3-triazole-quinoline against Plasmodium falciparum. *Chem. Biol. Drug Des.* **2014**, *84*, 325–332. [CrossRef] [PubMed]

10. Raj, R.; Singh, P.; Singh, P.; Gut, J.; Rosenthal, P.J.; Kumar, V. Azide-alkyne cycloaddition en route to 1H-1,2,3-triazole-tethered 7-chloroquinoline-isatin chimeras: Synthesis and antimalarial evaluation. *Eur. J. Med. Chem.* **2013**, *62*, 590–596. [CrossRef] [PubMed]

11. Melato, S.; Coghi, P.; Basilico, N.; Prosperi, D.; Monti, D. Novel 4-Aminoquinolines through Microwave-Assisted S$_N$Ar Reactions: A Practical Route to Antimalarial Agents. *Eur. J. Org. Chem.* **2007**, *36*, 6118–6123. [CrossRef]

12. De Souza, M.V.; Pais, K.C.; Kaiser, C.R.; Peralta, M.A.; de L Ferreira, M.; Lourenço, M.C. Synthesis and in vitro antitubercular activity of a series of quinoline derivatives. *Bioorg. Med. Chem.* **2009**, *174*, 1474–1480. [CrossRef] [PubMed]

13. Feldman, A.K.; Colasson, B.; Fokin, V.V. One-Pot Synthesis of 1,4-Disubstitued 1,2,3-Triazole from In Situ Generated Azides. *Org. Lett.* **2004**, *6*, 3897–3899. [CrossRef] [PubMed]

14. Lipinski, C.A. Drug-like properties and the causes of poor solubility and poor permeability. *J. Pharmacol. Toxicol. Methods* **2000**, *44*, 235–249. [CrossRef]

15. Daina, A.; Zoete, V. A boiled-egg to predict gastrointestinal absorption and brain penetration of small molecules. *Chem. Med. Chem.* **2016**, *11*, 1117–1121. [CrossRef] [PubMed]

16. Hongbin, Y.; Chaofeng, L.; Lixia, S.; Jie, L.; Yingchun, C.; Zhuang, W.; Weihua, L.; Guixia, L.; Yun, T. AdmetSAR 2.0: Web-service for prediction and optimization of chemical ADMET properties. *Bioinformatics* **2018**, *35*, 1067–1069.

17. Guantai, E.M.; Ncokazi, K.; Egan, T.J.; Gut, J.; Rosenthal, P.J.; Smith, P.J.; Chibale, K. Design, synthesis, and in vitro antimalarial evaluation of triazole linked chalcone and dienone hybrid compounds. *Bioorg. Med. Chem.* **2010**, *18*, 8243–8256. [CrossRef] [PubMed]

18. Wang, Y.; Ji, K.; Lan, S.; Zhang, L. Rapid access to chroman-3-ones through gold-catalyzed oxidation of propargyl aryl ethers. *Angew. Chem. Int. Ed.* **2012**, *51*, 1915–1918. [CrossRef] [PubMed]

19. Tiwari, M.K.; Coghi, P.; Agarwal, P.; Shyamlal, R.B.K.; Yadav, L.; Sharma, R.; Yadav, D.K.; Sahal, D.; Wong, V.K.W.; Chaudhary, S. Novel functionalized 1,2,4-Trioxanes as Potent Antimalarial and Anticancer Agents: Design, Synthesis, Structure Activity Relationship and in silico docking studies. *Chem. Med. Chem.* **2020**, *15*, 1216. [CrossRef] [PubMed]

molbank

MDPI

Short Note

Kokosanolide D: A New Tetranortriterpenoid from Fruit Peels of *Lansium domesticum* Corr. cv Kokossan

Fawwaz M. Fauzi, Sylvia R. Meilanie, Zulfikar, Kindi Farabi, Tati Herlina, Jamaludin Al Anshori and Tri Mayanti *

Departement of Chemistry, Faculty of Mathematics and Natural Sciences, Universitas Padjadjaran, Jatinangor 45363, Indonesia; fawwaz18002@mail.unpad.ac.id (F.M.F.); sylvia14001@mail.unpad.ac.id (S.R.M.); zulfikar15002@mail.unpad.ac.id (Z.); kindi.farabi@unpad.ac.id (K.F.); tati.herlina@unpad.ac.id (T.H.); jamaludin.al.anshori@unpad.ac.id (J.A.A.)
* Correspondence: t.mayanti@unpad.ac.id; Tel.: +62-8132-010-2633

Abstract: A novel tetranortriterpenoid named kokosanolide D has been isolated from fruit peels of *Lansium domesticum*. The structure of kokosanolide D was elucidated primarily on the basis of spectroscopic data including infrared, 1D and 2D-NMR, as well as high resolution mass spectroscopy analysis and comparison with related compounds previously reported.

Keywords: kokosanolide; tetranortriterpenoid; *Lansium domesticum*; Meliaceae

1. Introduction

Citation: Fauzi, F.M.; Meilanie, S.R.; Zulfikar; Farabi, K.; Herlina, T.; Al Anshori, J.; Mayanti, T. Kokosanolide D: A New Tetranortriterpenoid from Fruit Peels of *Lansium domesticum* Corr. cv Kokossan. *Molbank* **2021**, *2021*, M1232. https://doi.org/10.3390/M1232

Academic Editor: Giovanni Ribaudo

Received: 9 May 2021
Accepted: 5 June 2021
Published: 10 June 2021

Publisher's Note: MDPI stays neutral with regard to jurisdictional claims in published maps and institutional affiliations.

Lansium domesticum Corr. (Meliaceae) is a tree that grows widely in Indonesia and various countries in Southeast Asia [1,2]. In addition, this tree is also found in Australia, Suriname and Puerto Rico [3]. *Lansium domesticum* Corr. cv *Kokossan*, which is commonly called "kokosan" in Indonesia, thrives and bears fruit during the rainy season [4]. Several types of triterpenoid compounds have been reported from *L. domesticum* Corr., such as tetranortriterpenoid [5], triterpenoid glycosides [6,7], onoceranoids [4,8–10] and cycloartane triterpenoids [11], which showed various biological activities, such as being antifeedant [12], anticancer [13], antibacterial [14], antimutagenic [15,16], and antimalarial [17].

During our previous research on triterpenoid compounds from Indonesian Meliaceae plants, we have isolated two tetranortriterpenoid compounds (kokosanolide A and C) from the fruit seeds and one onoceranoid type triterpenoid compound (kokosanolide B) from the bark of *L. domesticum* cv *Kokossan* [12]. In the present study, we isolated a new tetranortriterpenoid compound from the methanol extract of fruit peels of *L. domesticum* Corr. cv *Kokossan* which is named kokosanolide D (**1**) (Figure 1). The chemical structure of compound **1** is determined by spectroscopic data, including infrared, 1D and 2D-NMR and HRMS.

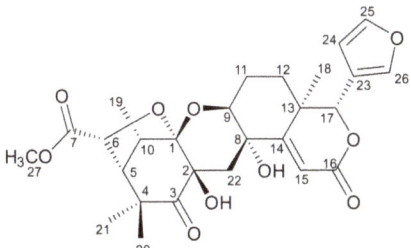

Figure 1. Chemical Structure of Compound **1**.

2. Results

Extraction and Isolation

The dried fruit peels of *L. domesticum* (1.7 Kg) was macerated at room temperature with *n*-hexane (5 × 2 L), ethyl acetate (5 × 2 L), and methanol (5 × 2 L). The solvents were removed by evaporation to give a crude *n*-hexane extract (86 g), ethyl acetate (110 g), and methanol (75 g). The methanol extract (75 g) was partitioned using butanol:H_2O (1:1) to give the butanol fraction (24 g). The butanol fraction (24 g) was fractionated by vacuum liquid chromatography on silica gel using a 10% gradient of *n*-hexane-ethyl acetate-methanol to give seventeen fractions (A–Q). Fraction G–H (2.9 g) was separated by silica gel open column chromatography using a 5% gradient of dichloromethane-ethyl acetate to give thirty-four fractions (GH1–GH34). Fraction GH13–15 (231.0 mg) was separated with silica gel open column chromatography using a 0.5–1% gradient of chloroform-acetone to give compound **1** (14.2 mg).

Kokosanolid D, colorless amorphous powder; HR-TOFMS *m*/*z* 517.2075 [M + H]$^+$ (calcd. for $C_{27}H_{33}O_{10}$, *m/z* 517,2074); IR (KBr) v_{max} (cm^{-1}): 3441, 1760, 1726, 1698, 1440, 1392; ^{1}H-NMR (CDCl$_3$, 500 MHz) and ^{13}C-NMR (CDCl$_3$, 125 MHz) shown in Table 1.

Table 1. ^{13}C-NMR, ^{1}H NMR and HMBC Spectral Data of Compound **1** in CDCl$_3$.

Position	δ_C	δ_H (ΣH, mult., *J* (Hz))	HMBC
1	107.0	-	-
2	77.1	-	-
3	209.7	-	-
4	47.8	-	-
5	55.9	2.30 (1H, *d*, 4)	1, 3
6	76.7	4.82 (1H, *d*, 3.5)	7, 4
7	172.0	-	-
8	67.8	-	-
9	73.4	4.16 (1H, *d*, 2.5)	8, 12, 14
10	36.5	3.29 (1H, *d*, 7)	4, 5, 6, 19
11	20.9	1.75 (1H, *t*, 3.5) 2.18 (1H, *tt*, 6)	9, 13
12	26.6	1.22 (1H, *t*, 13) 1.71 (1H, *dd*, 5)	13
13	38.4	-	-
14	167.5	-	-
15	117.4	6.46 (1H, *s*)	8, 13, 16
16	165.7	-	-
17	81.8	5.15 (1H, *s*)	12, 13, 14, 18, 23, 24, 26
18	19.7	1.26 (3H, *s*)	12, 13, 14, 17
19	11.8	1.16 (3H, *d*, 6.5)	1, 5, 10
20	23.1	0.98 (3H, *s*)	3, 4, 5, 21
21	29.6	1.37 (3H, *s*)	3, 4, 5, 20
22	34.4	2.40 (1H, *d*, 15) 3.02 (1H, *d*, 15)	1, 2, 8, 9, 14
23	119.7	-	-
24	110.2	6.43 (1H, *s*)	23, 25
25	142.8	7.39 (1H, *s*)	23, 24
26	141.5	7.48 (1H, *s*)	23, 24
27	52.5	3.67 (3H, *s*)	7

3. Discussion

Kokosanolide D (**1**) is obtained as colorless and amorphous from chloroform acetone. The molecular formula is determined to be $C_{27}H_{32}O_{10}$ by LC-ESI-MS data (*m/z* 517.2076, [M + H]$^+$), which is combined with ^{1}H and ^{13}C-NMR spectral data (Table 1) with twelve degrees of unsaturation. The IR (KBr) spectrum shows bands which can be assumed to be derived from the hydroxyl groups (v_{max}: 3441 cm^{-1}), ketones (v_{max}: 1760 cm^{-1}), unsaturated ketones (v_{max}: 1726 cm^{-1}) isolated double bonds (v_{max}: 1698 cm^{-1}) and *gem*-dimethyl (v_{max}: 1440 cm^{-1} and 1392 cm^{-1}).

^{1}H-NMR (Figure S1) showed three singlet signals (δ 0.98, 1.26, and 1.37) from the tertiary methyl group and one doublet from the secondary methyl group at δ 1.16 which correlated with H-10 (δ 3.29, 1H, *d*, 7). The singlet signal also appears in the lower field area (δ 3.67), which is thought to be the C-27 methoxy proton signal. A more refined analysis of the ^{1}H-NMR spectrum reveals the signal characteristics of the tetranortriterpenoid skeleton in the presence of β-substituted furan signals (δ 7.48, 7.39 and 6.43) and olefinic signals of α, β-unsaturated ketone (δ 6.46, 1H, *s*).

^{13}C-NMR, DEPT and HMQC spectrum (Figures S2–S4) shows 27 carbon signals referring to the signal characteristics of the furan ring (δ 142.8, 141.5, 119.7, and 110.2), ketone (δ 209.7), two ester groups (δ 172.0 and 165.7), one carbon oxygenated by two oxygens (δ 107.0) and α, β-unsaturated ketone (δ 167.5 and 117.4), thus showing that compound **1** has a hexacyclic structure with the presence of furan groups. ^{1}H-^{1}H COSY (Figure S5) shows the proton correlation of H$_6$/H$_5$, H$_9$/H$_{11}$, H$_{10}$/H$_{19}$, H$_{11}$/H$_{12}$, H$_{21}$/H$_{20}$, H$_{25}$/H$_{24}$ and H$_{26}$/H$_{24}$. These correlations indicate the presence of a tetranortriterpenoid skeleton with a furan ring [17]. The correlation of the partial structure was further explained by the HMBC correlation spectral data (Figures 2 and S6).

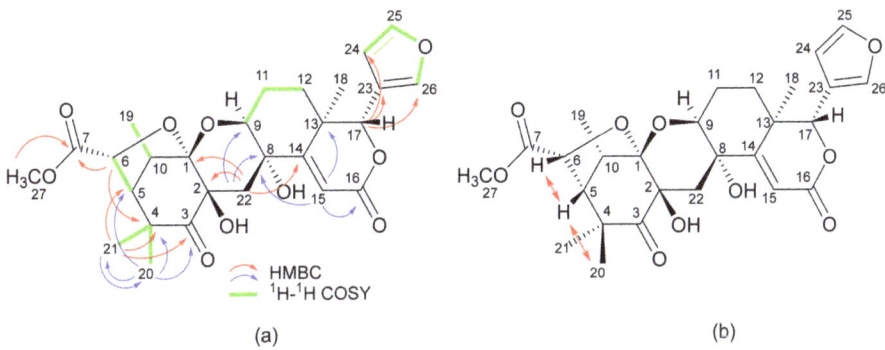

Figure 2. (a) ^{1}H-^{1}H COSY and HMBC and (b) NOESY correlations for compound **1**.

The correlation of oxygenated H-17 (δ 5.17) to C-23 (δ 119.7), C-24 (δ 110.2) and C-26 (δ 141.5) shows that C-17 binds to the furan ring. The pyran ring position is confirmed by the correlations of H-22 (δ 2.40 and 3.02) with C-1 (δ 107.0), C-2 (δ 77.1), C-8 (δ 67.8), C-9 (δ 73.4) and C-14 (δ 167.5). An α,β-unsaturated δ-lactone ring system is determined by the correlation of H-15 (δ 6.46) which connected with C-8 (δ 67.8), C-13 (δ 38.4) and C16 (δ 165.7). The correlations arising from two methyl (δ 0.98 and 1.37) to C-3 (δ 209.7), C-4 (δ 47.8) and C-5 (δ 55.9) indicate that *gem*-dimethyl is bound to C-4. Another correlation between the carbomethoxyl (δ 3.67) and H-6 (δ 4.82) signals to the carbonyl ester (δ 172.0) determines the position of the ester group attached to C-6. Basically, the NMR data of compound **1** were similar to those of kokosanolide A [12]. The only difference lies in C-8. Kokosanolide A has C-8 (δ 34.6), which is a metine, while in compound **1**, C-8 (δ 67.8) is a quaternary carbon that binds the hydroxyl group. The absence of a C-8 correlation with any protons in the HMQC data supports this suggestion. A NOESY spectrum shows H-5/ H-6/H20 correlation. This NOESY correlation is also found in kokosanolide A [12]. It can be assumed that kokosanolide A and compound **1** have the same stereochemical configuration and indicated that hydroxyl group C-8 should have α-orientation. Thus, the structure of compound **1** is determined to be a new tetranortriterpenoid from kokosanolide group named kokosanolide D.

4. Materials and Methods

4.1. General Experimental Procedures

Mass spectra were measured with a waters Xevo QTOFMS (Waters, Milford, MA, USA). IR spectra were measure on a One Perkin Elmer infrared-100. NMR data were recorded on a JNM-ECZ500R/S1 spectrometer at 500 MHz for ^{1}H and 125 MHz for ^{13}C using TMS as an internal standard. Chromatographic separations were carried out on silica gel G60 (Merck, Darmstadt, Germany) and RP18 (Merck). TLC plates were pre-coated with silica gel GF254 (Merck, 0.25 mm), and detection was achieved by spraying with 10% H_2SO_4 in ethanol, followed by heating.

4.2. Plant Material

The fruit peels of *L. domesticum* were collected from Cililin, West Java, Indonesia in April 2018. The plant was identified and deposited in the Herbarium Laboratory of the Department of Biology, Faculty of Mathematics and Natural Sciences, Universitas Padjadjaran, Indonesia (Identification Number: 195/HB/08/2018).

5. Conclusions

A new tetranortriterpenoid was isolated from the methanol extract of fruit peels of *L. domesticum* Corr. cv *kokossan*, which is named kokosanolide D.

Supplementary Materials: The following are available online, Figure S1. 1H-NMR spectrum of 1 (500 MHz in CDCl$_3$), Figure S2. 13C-NMR spectrum of 1 (125 MHz in CDCl$_3$), Figure S3. DEPT_135° spectrum of 1 (125 MHz in CDCl$_3$), Figure S4. HMQC spectrum of 1, Figure S5. 1H-1H COSY spectrum of 1, Figure S6. HMBC spectrum of 1, Figure S7. NOESY spectrum of 1, Figure S8. Infrared spectrum of 1 (in KBr), Figure S9. HR-TOF-MS spectrum of 1, Figure S10. TLC profile of 1.

Author Contributions: The following statements should be used Conceptualization, F.M.F. and T.M.; methodology, F.M.F., T.H.; software, F.M.F. and Z.; validation, K.F., T.M. and T.H.; formal analysis, F.M.F. and S.R.M.; investigation, T.M. and J.A.A.; resources, T.M.; data curation, K.F. and J.A.A.; writing—original draft preparation, F.M.F.; writing—review and editing, T.M. and J.A.A.; visualization, S.R.M. and T.H.; supervision, T.M. and T.H.; project administration, S.R.M.; funding acquisition, T.M. All authors have read and agreed to the published version of the manuscript.

Funding: This research was funded by PD, Ministry of Research, Technology and Higher Education, Indonesia, grant number 1827/UN6.3.1/LT/2020 (T.M).

Acknowledgments: We thank Joko Kusmoro, M.P. at Jatinangor Herbarium for identification of the plant material, Ahmad Darmawan and Sofa Fajriah at the Research Center for Chemistry, Indonesian Science Institute, for performing the NMR measurements, Kansy Haikal at the Center Laboratory of Universitas Padjadjaran for performing the HR-TOF-MS measurements.

Conflicts of Interest: The authors declare no conflict of interest.

References

1. Kiang, A.K.; Tan, E.L.; Lim, F.Y.; Habaguchi, K.; Nakanishi, K.; Fachan, L.; Ourisson, G. Lansic acid, a bicyclic triterpene. *Tetrahedron Lett.* **1967**, *37*, 3571–3574. [CrossRef]
2. Ghani, U. Chapter four—Terpenoids and steroids. In *Alpha-Glucosidase Inhibitors, Clinically Promising Candidates for Anti-Diabetic Drug Discovery*; Elsevier: Amsterdam, The Netherlands, 2020; pp. 101–117.
3. Techavuthiporn, C. Langsat—Lansium *domesticum*. In *Exotic Fruits Reference Guides*; Rodrigues, S., de Brito, E., Silva, E., Eds.; Academic Press: London, UK, 2018; pp. 279–283.
4. Dong, S.H.; Zhang, C.R.; Dong, L.; Wu, Y.; Yue, J.M. Onoceranoid-type triterpenoids from *Lansium domesticum*. *J. Nat. Prod.* **2011**, *74*, 1042–1048. [CrossRef] [PubMed]
5. Mayanti, T.; Supratman, U.; Mukhtar, M.R.; Awang, K.; Ng, S.W. Kokosanolide from the seed of *Lansium domesticum* Corr. *Acta Crystallogr.* **2009**, *E65*, o750.
6. Nishizawa, M.; Nishide, H.; Hayashi, Y.; Kosela, S. The structure of lansioside A: A novel triterpene glycoside with amino-sugar from *Lansium domesticum*. *Tetrahedron Lett.* **1982**, *23*, 1349–1350. [CrossRef]
7. Nishizawa, M.; Nishide, H.; Kosela, S.; Hayashi, Y. Structure of lansiosides: Biologically active new triterpene glycosides from *Lansium domesticum*. *J. Org. Chem.* **1983**, *48*, 4462–4466. [CrossRef]

8. Tanaka, T.; Ishibashi, M.; Fujimoto, H.; Okuyama, E.; Koyano, T.; Kowiyhayakorn, T.; Hayashi, M.; Komiyama, K. New onoceranoid constituents from *Lansium domesticum*. *J. Nat. Prod.* **2002**, *65*, 1709–1711. [CrossRef] [PubMed]
9. Habaguchi, K.; Watanabe, M.; Nakadaira, Y.; Nakanishi, K.; Kaing, A.K.; Lim, F.L. The full structures of lansic acid and its minor congener, an unsymmetric onoceradienedione. *Tetrahedron Lett.* **1986**, *34*, 3731–3734. [CrossRef]
10. Zulfikar; Putri, N.K.; Fajriah, S.; Yusuf, M.; Maharani, R.; Al Anshori, J.; Supratman, U.; Mayanti, T. 3-Hydroxy-8,14-secogammacera-7,14-dien-21-one: A new onoceranoid triterpenes from *Lansium domesticum* Corr. cv kokossan. *Molbank* **2020**, *2020*, M1157. [CrossRef]
11. Mayanti, T.; Sianturi, J.; Harneti, D.; Darwati; Supratman, U.; Rosli, M.M.; Fun, H.K. 9,19-Cyclolanost-24-en-3-one,21,23-epoxy-21,22-dihydroxy (21*R*, 22*S*, 23*S*) from the leaves of *Lansium domesticum* Corr cv Kokossan. *Molbank* **2015**, *2015*, M880. [CrossRef]
12. Mayanti, T.; Tjokronegoro, R.; Supratman, U.; Mukhtar, M.R.; Awang, K.; Hadi, A.H.A. Antifeedant Triterpenoids from the Seeds and Bark of *Lansium domesticum* cv Kokossan (Meliaceae). *Molecules* **2011**, *16*, 2785–2795. [CrossRef] [PubMed]
13. Fadhilah, K.; Wahyuona, S.; Astuti, P. A bioactive compound isolated from duku (*Lansium domesticum* Corr) fruit peels exhibits cytotoxicity against T47D cell line. *F1000Research* **2020**, *9*, 3. [CrossRef]
14. Ragasa, C.Y.; Labrador, P.; Rideout, J.A. Antimicrobial terpenoid from *Lansium domesticum*. *Philipp. Agric. Sci.* **2006**, *89*, 101–105.
15. Matsumoto, T.; Kitagawa, T.; Teo, S.; Anai, Y.; Ikeda, R.; Imahori, D.; Ahmad, H.S.; Watanabe, T. Structures and antimutagenic effects of onoceranoid-type triterpenoids from the leaves of *Lansium domesticum*. *J. Nat. Prod.* **2018**, *81*, 2187–2194. [CrossRef] [PubMed]
16. Matsumoto, T.; Kitagawa, T.; Ohta, T.; Yoshida, T.; Imahori, D.; Teo, S.; Ahmad, H.S.; Watanabe, T. Structures of triterpenoids from the leaves of *Lansium domesticum*. *J. Nat. Med.* **2019**, *73*, 727–734. [CrossRef] [PubMed]
17. Saewan, N.; Sutherland, J.D.; Chantrapromma, K. Antimalarial tetranortriterpenoids from the seed of *Lansium domesticum* Corr. *Phytochemistry* **2006**, *67*, 2288–2293. [CrossRef] [PubMed]

MDPI

Communication

Bioactive Antidiabetic Flavonoids from the Stem Bark of *Cordia dichotoma* Forst.: Identification, Docking and ADMET Studies

Nazim Hussain [1], Bibhuti Bhushan Kakoti [2], Mithun Rudrapal [3,*], Khomendra Kumar Sarwa [4], Ismail Celik [5], Emmanuel Ifeanyi Attah [6], Shubham Jagadish Khairnar [7], Soumya Bhattacharya [8], Ranjan Kumar Sahoo [9] and Sanjay G. Walode [3]

[1] Kingston Imperial Institute of Technology and Science, Dunga, Dehradun 248007, India; nhussain116@gmail.com
[2] Department of Pharmaceutical Sciences, Dibrugarh University, Dibrugarh 786004, India; bibhutikakoti@dibru.ac.in
[3] Rasiklal M. Dhariwal Institute of Pharmaceutical Education & Research, Chinchwad, Pune 411019, India; sanjuwalode@rediffmail.com
[4] Department of Pharmacy, Government Girls Polytechnic, Raipur 492001, India; khomendra.sarwa@gmail.com
[5] Department of Pharmaceutical Chemistry, Faculty of Pharmacy, Erciyes University, Talas, Kayseri 38280, Turkey; celikismail66@gmail.com
[6] Department of Pharmaceutical and Medicinal Chemistry, University of Nigeria, Nsukka 410001, Nigeria; emmanuel.attah.pg00429@unn.edu.ng
[7] MET Institute of Pharmacy, Bhujbal Knowledge City, Nashik 422003, India; sunnykhairnar62@gmail.com
[8] Guru Nanak Institute of Pharmaceutical Science and Technology, 157/F, Nilgunj Road, Panihati, Kolkata 700114, India; soumya.bhattacharya@gnipst.ac.in
[9] School of Pharmacy and Life Sciences, Centurion University of Technology and Management, Bhubaneswar 752050, India; ranjankumar.sahoo@cutm.ac.in
* Correspondence: rsmrpal@gmail.com; Tel.: +91-86-3872-4949

Citation: Hussain, N.; Kakoti, B.B.; Rudrapal, M.; Sarwa, K.K.; Celik, I.; Attah, E.I.; Khairnar, S.J.; Bhattacharya, S.; Sahoo, R.K.; Walode, S.G. Bioactive Antidiabetic Flavonoids from the Stem Bark of *Cordia dichotoma* Forst.: Identification, Docking and ADMET Studies. *Molbank* 2021, 2021, M1234. https://doi.org/10.3390/M1234

Academic Editor: Giovanni Ribaudo

Received: 7 May 2021
Accepted: 9 June 2021
Published: 11 June 2021

Publisher's Note: MDPI stays neutral with regard to jurisdictional claims in published maps and institutional affiliations.

Abstract: *Cordia dichotoma* Forst. (F. Boraginaceae) has been traditionally used for the management of a variety of human ailments. In our earlier work, the antidiabetic activity of methanolic bark extract of *C. dichotoma* (MECD) has been reported. In this paper, two flavonoid molecules were isolated (by column chromatography) and identified (by IR, NMR and mass spectroscopy/spectrometry) from the MECD with an aim to investigate their antidiabetic effectiveness. Molecular docking and ADMET studies were carried out using AutoDock Vina software and Swiss ADME online tool, respectively. The isolated flavonoids were identified as 3,5,7,3',4'-tetrahydroxy-4-methoxyflavone-3-*O*-L-rhamnopyranoside and 5,7,3'-trihydroxy-4-methoxyflavone-7-*O*-L-rhamnopyranoside (quercitrin). Docking and ADMET studies revealed the promising binding affinity of flavonoid molecules for human lysosomal α-glucosidase and human pancreatic α-amylase with acceptable ADMET properties. Based on computational studies, our study reports the antidiabetic potential of the isolated flavonoids with predictive pharmacokinetics profile.

Keywords: *C. dichotoma*; flavonoids; antidiabetic; α-glucosidase; α-amylase; docking; ADMET

1. Introduction

Cordia dichotoma Forst. (also known as Indian cherry, F. Boraginaceae) is a traditionally important deciduous medicinal plant widely grown in India, Sri Lanka and other tropical countries of the world [1]. This plant has been traditionally (Ayurveda, Unani and Siddha medicines) used for the management of a variety of human ailments/disorders [1]. Leaves and stem bark have been used traditionally in the treatment of fever, dyspepsia, diarrhea, leprosy, gonorrhea and wounds [2]. Leaves, seeds, bark and fruits have been reported to

39

exhibit anti-inflammatory, anthelmintic, antibacterial, antileprotic, antiviral, diuretic, astringent, demulcent, laxative/purgative, expectorant/antitussive, tonic, immunomodulatory, hepatoprotective and gastroprotective/antiulcer activities [3,4].

Recent literature have also reported the anti-inflammatory [3], antidiabetic [4], anticancer [2] and antioxidant [3] activities for the bark extract of *C. dichotoma*. Phytoconstituents like3′,5-dihydroxy-4′-methoxyflavanone-7-*O*-α-L-rhamnopyranoside, β-sitosteryl-3β-glucopyranoside-6′-*O*-palmitate, quercitrin and β-sitosterol have already been isolated from the leaves of this plant [5]. Betulin, lupeol-3-rhamnoside, β-sitosterol, taxifolin-3-5-dirhamnoside, hesperitin-7-rhamnoside, rutin, chlorogenic acid and caffeic acid have been reported from seeds of *C. dichotoma* [2–5]. Flavonoids, the most important class of plant polyphenolics possessing diverse range of biological/pharmacological potential [6–8], are attributed to be the most dominant phytochemical components in various plant parts of *C. dichotoma*. In this work, the isolation and identification of bioactive flavonoids from the methanolic bark extract of *C. dichotoma* Forst was carried out. The isolated flavonoid molecules were further investigated for their antidiabetic potential and pharmacokinetic properties by molecular docking and ADMET studies.

2. Results and Discussion

The phytochemical analysis revealed the presence of flavonoids, alkaloids, glycosides, saponins, steroids, carbohydrates and proteins in the methanolic bark extract of *C. dichotoma* Forst (MECD) [2–4].

2.1. Identification of Isolated Phytocompounds
2.1.1. Compound 1 (MECD-1)

Subfraction 20–78 was purified by column chromatography on silica using methanol: ethyl acetoacetate to yield the pure compound 1 (120 mg). The isolated compound 1 was obtained as pale yellow amorphous powder with a melting range of 163–165 °C and a $[\alpha]^{25}_D$ value of +0.34 (conc. 0.25 mg/mL, MeOH). The structure of the isolated compound 1, represented in Figure 1, was elucidated by UV, IR, ^{1}H-NMR, ^{13}C-NMR and mass spectroscopic/spectrometric analyses (Figure S1A–D).

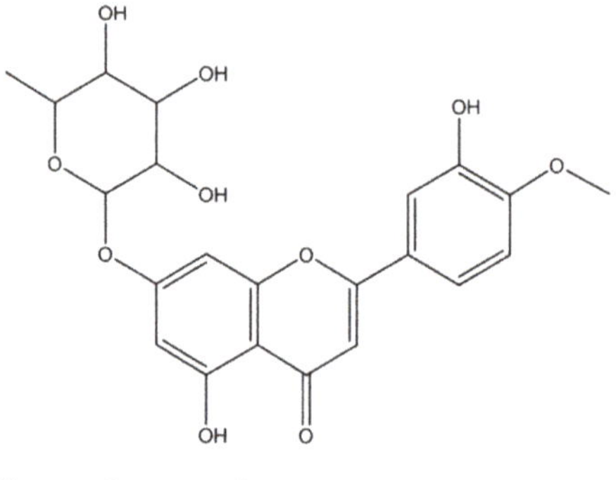

Figure 1. Structure of compound 1 (MECD-1).

UV (λ_{max}, nm, MeOH): 307, 334; FT-IR (υ_{max}, cm^{-1}, KBR): 3265, 2960, 2895, 1668, 1616, 1429, 1367; ^{1}H-NMR (δ, 400 MHz, DMSO-d6): 0.79 (J = 6.0 Hz), 3.98 (J = 15.0, 3.5 Hz), 2.96 (J = 15.0, 11.0 Hz), 3.02 (J = 11.0, 3.5 Hz), 4.41, 4.43, 5.25, 6.21, 6.42 (J = 2.0 Hz), 7.26 (J = 6.5, 2.0 Hz); ^{13}C-NMR (δ, 100 MHz, DMSO-d6): 193.56, 102.25, 55.78, 52.55.

The IR bands at 3265 cm^{-1} (O-H stretching) revealed the presence of the hydroxyl group in the structure of compound 1. Other prominent absorption bands at 2960 and 2895 cm^{-1} (aliphatic C-H stretching), 1668 cm^{-1} (C=O stretching) and 1616 cm^{-1} (aromatic C=H stretching) indicated the presence of methyl group (CH$_3$), α,β unsaturated carbonyl group and aromatic rings. In ^{1}H-NMR spectrum, three one-proton double doublets at δ 3.98 (J = 15.0, 3.5 Hz), 3.02 (J = 11.0, 3.5 Hz) and 2.96 (J = 15.0, 11.0 Hz) were ascribed to H-2, H-3α and H-3β protons, respectively of ring C of a flavone moiety. Two one-proton doublets at δ 6.21 and 6.42 (J = 2.0 Hz, each) were assigned to H-6 and H-8 aromatic protons. Two doublets at δ 7.26 (J = 6.5, 2.0 Hz), each integrating for one proton, were ascribed correspondingly to H-2, H-5 and H-6 of aromatic protons. A three-proton singlet at δ 3.61 was attributed to methoxy (OCH$_3$) protons. A broad singlet at δ 5.25 was accounted to H-1″ anomeric proton, while a three proton doublet at δ 0.79 (J = 6.0 Hz) was appeared due to secondary methyl proton H-6″ of rhamnose unit. The remaining protons of sugar unit appeared between δ 4.43 and 4.41. The ^{13}C-NMR spectra showed twenty two distinct signals suggesting that the compound contains twenty two carbon atoms. The important signals appeared at δ 193.56 (C-4, carbonyl carbon), 102.25 (C-1 anomeric carbon), 55.78 (methoxy carbon OCH$_3$) and 17.95 (C-6″ methyl carbon). The presence of an aromatic methoxy group was confirmed by position of the methyl signal at δ 52.55. The molecular ion [M]$^+$ peak was obtained at *m/z* 446.0, which concord the molecular formula of the compound 1 as C$_{22}$H$_{22}$O$_{11}$. The NMR spectral data also supported the structure of the compound. A thorough spectral interpretation suggests that the compound 1 (MECD-1) is **5,7,3,-trihydroxy-4-methoxyflavone-7-*O*-L-rhamnopyranoside**.

2.1.2. Compound 2 (MECD-2)

Subfraction 120–184 was purified by column chromatography on silica gel using methanol:ethyl acetoacetate to obtain the pure compound 1 (160 mg). The isolated compound 1 was obtained as pale yellow crystalline powder with a melting range of 176–178 °C and a [α]^{25}$_D$ value of +0.36 (conc. 0.25 mg/mL, MeOH). The structure of the isolated compound 1, represented in Figure 2 was elucidated by UV, IR, ^{1}H-NMR, ^{13}C-NMR and mass spectroscopic/spectrometric analyses (Figure S2A–D).

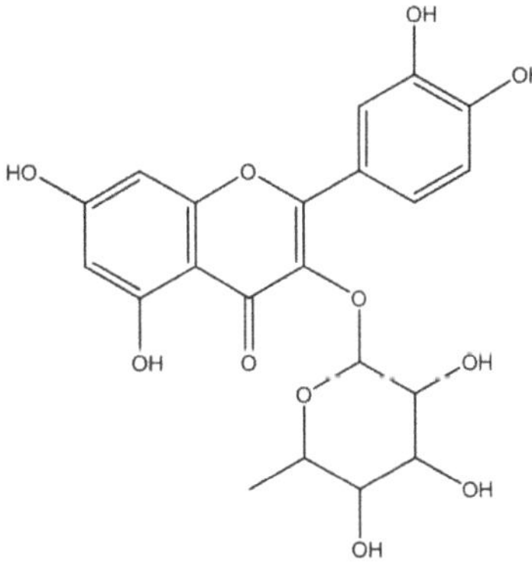

Figure 2. Structure of compound 2.

UV (λ$_{max}$, nm, MeOH): 312, 346; FT-IR (υ$_{max}$, cm^{-1}, KBR): 3265, 2950, 2880, 1654, 1502, 1454, 1354, 810. ^{1}H-NMR (δ, 400 MHz, DMSO-d$_6$): 0.73 (J = 7.1 Hz), 3.96–4.81, 6.11, 6.29 (J = 2.1 Hz), 6.78 (J = 9.6 Hz), 7.14 (J = 9.6, 2.2 Hz), 7.20 (J = 2.2 Hz), 12.55; ^{13}C-NMR (δ, 100 MHz, DMSO-d$_6$): 156.90 (C-2), 134.69 (C-3), 178.20 (C-4), 104.56 (C-4), 161.75 (C-5), 99.15(C-6), 164.63 (C-7), 64.09 (C-8), 157.76(C-8), 121.22 (C-1′), 116.13 (C-2′),145.65 (C-3′), 148.88 (C-4′), 2115.93 (C-5′), 121.58 (C-6′), 102.29 (C-1″), 71.03 (C-2″), 70.83(C-3″), 71.66 (C-4″), 70.51 (C-5″), 17.95 (C-6″).

The IR bands at 3265 cm^{-1} (O-H stretching) revealed the presence of hydroxyl group in the structure of compound 2. Other prominent absorption bands at 2950 and 2880 cm^{-1} (aliphatic C-H stretching), 1654 cm^{-1} (C=O stretching) and 1502 cm^{-1} (aromatic C=H stretching) indicated the presence of methyl group (CH$_3$), α,β unsaturated carbonyl group and aromatic rings. The ^{1}H-NMR spectrum exhibited a set of two coupled doublets at δ 6.11 and 6.29 (J = 2.1 Hz, each), which was ascribed to H-6 and H-8 aromatic protons. Another set of coupled signals consisting of two doublets at δ 7.20 (J = 2.2 Hz), 6.78 (J = 9.6 Hz) and a double-doublet at δ 7.14 (J = 9.6, 2.2 Hz) were ascribed to H-2′, H-5′ and H-6′ aromatic protons of ring B. A doublet at δ 5.15 (J = 8.1 Hz) was assigned to H-1″ anomeric proton, while as another doublet at δ 0.73 (J = 7.1 Hz) was attributed to methyl protons (H-6″) of rhamnose unit. The remaining protons of rhamnose resonated from δ 4.81 to 3.96. A single proton singlet at δ 12.55 was attributed to hydroxyl proton. The ^{13}C-NMR spectrum displayed signals for twenty-one carbons. Important signals appeared for carbonyl carbon (δ 178.20, C-4), anomeric carbon (δ 102.29, C-1″) and methyl carbon (δ 17.94, C-6″). The molecular ion [M]+ peak was obtained at *m/z* 448.0, which concord the molecular formula of the compound as $C_{21}H_{20}O_{11}$. The ^{1}H and ^{13}C NMR data was compared with other reported flavonoids and was found to be **3,5,7,3,4,-tetrahydroxy-4-methoxyflavone-3-*O*-L-rhamnopyranoside** (Quercitrin (Quercetin-3-*O*-L-rhamnoside).

2.2. Molecular Docking

Molecular docking is used to understand the drug–receptor interaction, binding affinity and binding orientation of bioactive molecules into the target protein molecule. The objective behind docking study is to predict a particular biological activity based on the binding orientation/affinity of small molecule ligands to the appropriate target active site [9]. In the docking study, the binding affinity was predicted in terms of the interaction energy (kcal/mole). Results of docking (binding) energies are given in Table 1. Both the compounds exhibited very good binding affinity against both α-glucosidase and α-amylase enzyme. Compound 1 (MECD-1) exhibited more binding affinity against alpha-amylase compared to the alpha-glucosidase. On the hand, the compound 2 (MECD-2) showed better affinity against alpha-glucosidase than alpha-amylase. Not much variation in binding energies between these two enzymes were observed. Docking scores of isolated compounds were compared with that of the standard drug, acarbose. Against α-glucosidase, the binding affinity of compounds 1 and 2 was comparatively more affinity than the standard drug. On the other hand, the binding affinity of compounds 1 and 2 were less to some extent than that the standard drug against α-amylase. No significant difference in activities between isolated test compounds and the standard compound was observed. The test compounds were found to have α-glucosidase and α-amylase inhibitory potential to a similar degree as that of the standard drug, acarbose. Overall, both the isolated flavonoids exhibited significant inhibitory potential of human glucosidase and amylase enzymes.

Table 1. Docking data of compounds.

Compound	Binding Energy (kcal/mole)	
	5NN8	1B2Y
Compound 1 (MECD-1)	−7.8	−8.6
Compound 2 (MECD-2)	−8.0	−7.8
Acarbose (Standard drug)	−7.6	−9.4

5NN8: Human lysosomal acid α-glucosidase; 1B2Y: Human pancreatic α-amylase; MECD: Metanolic bark extract of *C. dichotoma*.

Post-docking visualization of protein–ligand complexes revealed that the compounds interacted with active site residues of the protein molecules through the formation of predominantly hydrogen bonding interactions (Figures 3 and 4). From the observation of 2D interaction diagrams of compound 1-α-glucosidase complexes, it is clear that compound 1 formed H-bonds with Trp59, Gln63, Asp197, Asp300 and Asp356 residues, whereas the compound 2 interacted with Tyr62, His101, His201 and Gly306 residues through H-bonds (Figure 3a,b). The 3D diagrams revealed the binding conformation and binding poses of the compounds at the catalytic site of α-glucosidase (Figure 3c,d) were observed.

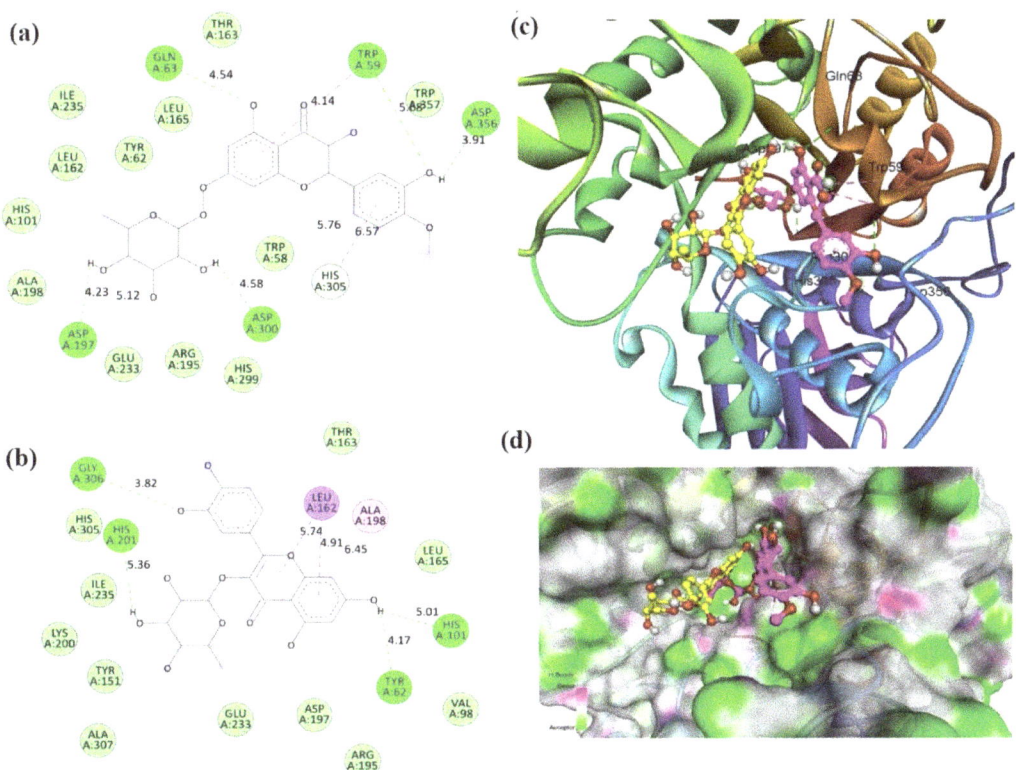

Figure 3. (**a**) Two-dimensional interactions between compound 1 and α-glucosidase, (**b**) 2D interactions between compound 2 and α-glucosidase showing hydrogen bonding and other non-covalent interactions with amino acid residues at the active site, (**c**) 3D representation of protein-ligand interactions showing binding conformation and (**d**) binding poses/binding modes of both the compounds at the catalytic site of α-glucosidase.

From the observation of 2D interaction diagrams of compound 2-α-amylase complexes, it is clear that compound 1 formed prominent H-bonds with Asp404, Ser523 and Ser524 residues, whereas the compound 2 interacted with Asp616, His674 and Leu678 residues through H-bonds (Figures 3b and 4a). The 3D diagrams revealed the binding conformation and binding poses of the compounds at the catalytic site of alpha-amylase (Figures 3d and 4c) were observed.

Upon critical analysis of protein–ligand interactions, favorable binding orientations and/or binding modes of flavonoid molecules for both the α-glucosidase and α-amylase enzyme were evident. Both the compounds structurally represent glycosides of flavones, which are abundantly found in plant kingdom. The glycone part (sugar), i.e., the rhamnose moiety is similar in both the compounds, while the aglycone part (non-sugar bioactive principle, flavanone moiety) is dissimilar. In compound 1, it is 5,7,3'-trihydroxy-4-methoxyflavone, whereas, it is 3,5,7,3',4'-tetrahydroxy-4-methoxyflavone, i.e., quercetin in compound 2. The aglycone, i.e., the flavonoid moiety is a polyhydroxylated C_6-C_3-C_6 tricyclic heteroaromatic system (phenylchromone) [10–12] with distinct structural features, particularly in terms of nature and pattern of ring substitutions. There is a close structural resemblance between these two isolated flavonoid glycosides. The basic flavone skeleton along with hydroxylated/methoxylated aromatic ring interacted predominantly with the active site residues of target protein molecules. Polar groups such as phenolic hydroxy groups and carbonyl moiety contributed significantly in protein-ligand interactions with the formation of hydrogen bonds. Apart from hydrogen bonding, other non-bonding interactions such as hydrophobic interactions also exist, but to a lesser extent. Aromatic bulky moieties chromone system and phenyl ring were mainly involved in non-polar hydrophobic interactions.

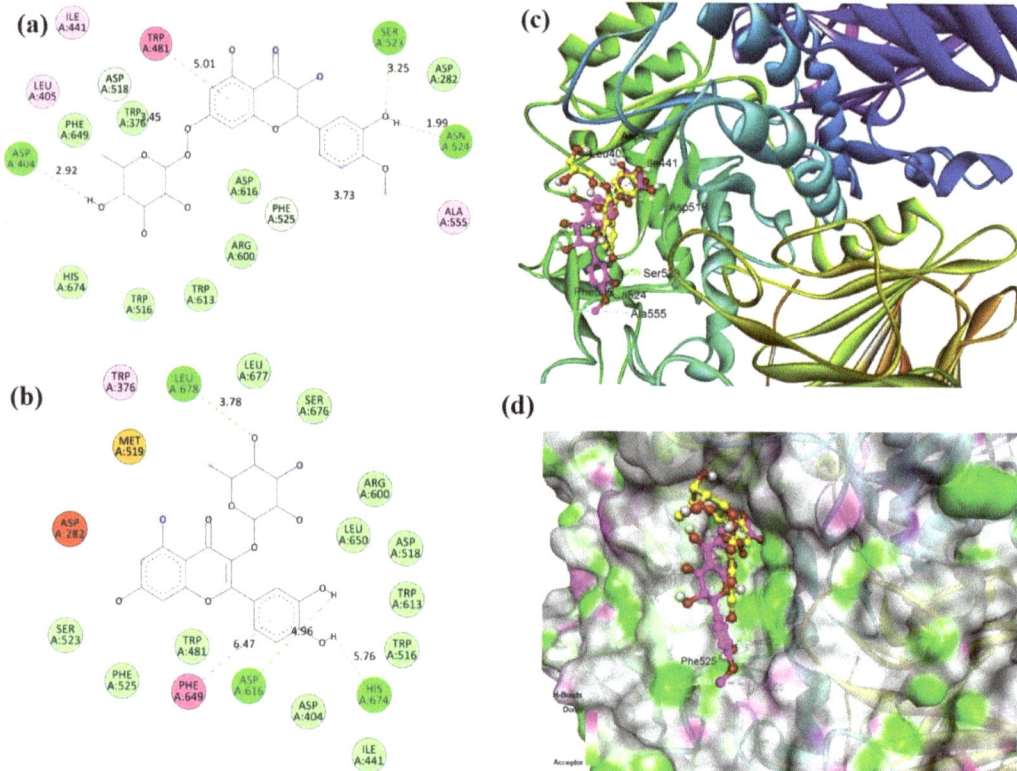

Figure 4. (**a**) Two-dimensional interactions between compound 1 and α-amylase, (**b**) 2D interactions between compound 2 and α-glucosidase showing hydrogen bonding and other non-covalent interactions with amino acid residues at the active site, (**c**) 3D representation of protein–ligand interactions showing binding conformation and (**d**) binding poses/binding modes of both the compounds at the catalytic site of α-amylase.

The human lysosomal α-glucosidase and pancreatic α-amylase enzymes play an important role in the digestion of dietary long-chain complex carbohydrates (breakdown of starch and disaccharides to glucose) and hence, their inhibition is believed to facilitate the reduction of post-prandial (post meal) blood glucose level in type 2 diabetes [13]. The traditional usefulness about the antidiabetic potential of *C. dichotoma* is mentioned in literature [2–4]. In our earlier studies, the antidiabetic activity of the methanolic bark extract of *C. dichotoma* has already been reported [4]. Moreover, the literature suggest that the flavonoids content demonstrates antidiabetic efficacy of many plants [14–18]. Our docking study validates the antidiabetic claim about *C. dichotoma* reported in traditional medicines and in recent literature. Although the isolated phytocompounds are already established bioactive flavonoids with many scientific reports from past literature, their antidiabetic potential determined by in silico (molecular docking) methods with α-glucosidase and α-amylase inhibitory activities has been reported for the first time. Our study may thus provide an avenue for further investigation with these bioflavonoids for their development as potent antidiabetic molecules with alpha-glucosidase and alpha-amylase inhibitory agents for the treatment of type 2 diabetes mellitus.

2.3. ADMET

Results of predicted ADMET (absorption, distribution, metabolism, excretion and toxicity) data showed that both the isolated compounds possess good solubility profile, which is in favor of their oral bioavailability. There is a prediction of poor intestinal absorption, while the compounds were predicted to be non-inhibitors of the cytochromes (CYP_{450}) [19]. Poor intestinal absorption might be due to their limited oil/water partition coefficient (logP) values (-1.64 and -1.84). CYP_{450} enzymes are largely involved in drug metabolism. Non-inhibition of CYP_{450} enzymes suggests that compounds do not suppress the metabolic function of the enzymes. Inhibition can lead to increased bioavailability of compounds that normally undergo extensive first-pass elimination or to decreased elimination of compounds dependent on metabolism for systemic clearance. Compounds did not exhibit the property of blood brain barrier (BBB) penetration. It substantiates that the compounds are devoid of producing CNS toxicities. Furthermore, quercitrin (compound 2) was predicted to be a substrate to permeability of glycoprotein (p-gp), whereas the other flavonoid molecule (compound 1) did not show such property. Glycoprotein is responsible for the efflux of drug molecules out of the target cells [20]. A good drug candidate should not only have sufficient efficacy against the therapeutic target, but also show appropriate ADMET properties at a therapeutic dose. It is therefore inevitable to evaluate the ADMET profile of drug-like molecules to avoid the failure of candidate drugs at the clinical stage of drug development [21].

3. Materials and Methods

3.1. Collection of Plant

The barks of *Cordia dichotoma* Forst. were collected from the Duhai forest of Ghaziabad, Uttar Pradesh, India. The plant species was identified from National Institute of Science Communication and Information Resources, New Delhi, India. The voucher specimen (NISCAIR/RHMD/Consult/2012-13/2025/33) of the bark of *Cordia dichotoma* Forst. was submitted at the herbarium of the department for future reference.

3.2. Preparation of Methanolic Bark Extract

The shade dried barks of *C. dichotoma* were pulverized to a coarse powder and defatted using petroleum ether by the cold maceration method [2] to remove fat, latex and non-polar compounds of high molecular weights. The defatted plant residues were then macerated successively with methanol to obtain the desired extract [3,4]. The collected extract was filtered through Whatman No. 1 filter paper. Rotary evaporator was used to concentrate the filtrate. The concentrated extract was preserved in refrigerator at 4 °C for further use. The percentage yield of the methanolic bark extract of *C. dichotoma* (MECD) was 7.11%.

3.3. Phytochemical Analysis

Chemical tests for the screening and detection of phytochemical constituents of the MECD were carried out using the standard procedures [22,23].

3.4. Isolation of Phytocompounds

The MECD was subjected to column chromatographic separation using silica gel (packed column, 100–200 mesh) and a glass column (6.0 × 3 inch dimension) [6] for the isolation of bioactive phytoconstituents in pure form. The elution was carried out by gradient separation technique using the solvent system of n-hexane/ethyl acetate. The column was eluted successively with n-hexane:ethyl acetate in increasing order of polarity (98:2, 95:5, 90:10, 80:20, 60:40, 50:50, 35:65, 30:70, 25:75, 20:80 and 100%). The fractions collected were subjected to thin-layer chromatography (TLC) to check their homogeneity. Chromatographically identical fractions (having the same R_f values) were combined together and concentrated. The concentrated fractions were purified by crystallization with methanol/benzene and confirmed by their sharp melting points.

3.5. Identification of Isolated Compounds

Ultraviolet (UV)–visible spectra were recorded on Shimadzu UV-1700 UV–visible spectrophotometer (Shimadzu, Kyoto, Japan) and the wave lengths of maximum absorption (λ_{max}, nm) were reported. Infrared (IR) spectra were obtained on a Bruker alpha Fourier transform (FT-IR) spectrometer (Bruker, MA, USA) using the KBR disc and reported in terms of frequency of absorption (υ_{max}, cm^{-1}). ^{1}H and ^{13}C nuclear magnetic resonance (NMR) spectra were recorded on Bruker Avance II 400 FT-NMR spectrometer (Bruker, MA, USA) at 400 and 100 MHz, respectively using tetramethylsilane (TMS) as an internal standard (δ 0.00 ppm) and CDCl$_3$ as a solvent. Mass spectra were obtained on a LC–MS Water 4000 ZQ instrument (Waters, Massachusetts, USA) using electrospray ionization (ES$^+$). The m/z values were recorded in the range of m/z between 100 and 500 and the m/z values of the most intense molecular ion [M]+ peak were considered. Melting points were determined on an electric melting point apparatus (JSGW, Model 3046). (Jain Scientific Glass Works, Ambala, India)

3.6. Molecular Docking

The X-ray crystal structure of proteins, viz., human lysosomal acid α-glucosidase (PDB ID: 5NN8) and human pancreatic α-amylase (PDB ID: 1B2Y) were reposited by Roig-Zamboni et al. [24] and Nahoun et al. [25] having resolution of 2.45 Å and 3.20 Å, respectively were retrieved from the RCSB protein data bank (http://www.rcsb.org/ (accessed on 13 March 2021)).

Prior to docking, The docking was performed in the AutoDock Vina software(The Scripps Research Institute, La Jolla, CA, USA) [26] in accordance with the standard procedure. The protein crystal structure was prepared prior to the docking process. Hydrogen atoms were added to the protein structure, and all ionizable residues were set at their default protonation at pH 7.4. The active site coordinates were determined with the dimensions of $x = -15.941$, $y = -37.643$, $z = 92.912$ for 5NN8 and $x = 22.116$, $y = 4.749$, $z = 45.878$ for 1B2Y and, a grid box with radius of $25 \times 25 \times 25$ A^3 was generated for both the proteins. Similarly, the ligands were prepared and energy minimized using Chem3D 17.0 software. During the docking process, the receptor was rigidly held, while the ligands were allowed to flex during the refinement. Binding energies of docking were recorded and analyzed. The best docked poses and binding modes of protein–ligand interactions were obtained using the Discovery Studio visualizer.

3.7. ADMET Prediction

Predictive ADMET (pharmacokinetics) parameters were studied using web-based Swiss ADME tool developed by Daina et al. [27]. Solubility, intestinal absorption, oil/water partition coefficient (logP), CYP$_{450}$ inhibition, blood brain barrier penetration and p-gp substrate were predicted [28].

4. Conclusions

This study reports two bioactive flavonoids, viz., 3,5,7,3′,4′-tetrahydroxy-4-methoxy-flavone-3-*O*-L-rhamnopyranoside and 5,7,3′-trihydroxy-4-methoxyflavone-7-*O*-L-rhamno-pyranoside (Quercitrin) isolated and identified from the methanolic bark extract *Cordia dichotoma* Forst. The molecular docking study investigated the antidiabetic potential of the isolated flavonoids against human lysosomal acid α-glucosidase and human pancreatic α-amylase enzymes. The predictive ADMET study demonstrated acceptable pharmacokinetics of the isolated compounds. The in silico study needs to be further validated by in vitro and in vivo experimental assays in order to confirm the antidiabetic effectiveness for the flavonoids reported herein. Our study may thus provide an avenue for further investigation with these bioflavonoids for their development as potent antidiabetic molecules with α-glucosidase and α-amylase inhibitory agents for the treatment of type 2 diabetes mellitus.

Supplementary Materials: The following are available online, Figure S1A: FT-IR spectrum of compound 1, Figure S1B: ^{1}H-NMR spectrum of compound 1, Figure S1C: ^{13}C-NMR spectrum of compound 1, Figure S1D: Mass spectrum of compound 1, Figure S2A: FT-IR spectrum of compound 2, Figure S2B: ^{1}H-NMR spectrum of compound 2, Figure S2C(1) and Figure S2C(2): ^{13}H-NMR spectrum of compound 2, Figure S2D: Mass spectrum of compound 2.

Author Contributions: Conceptualization, N.H. and B.B.K.; methodology, N.H.; software, I.C.; validation, I.C. and M.R.; formal analysis, M.R. and E.I.A.; investigation, N.H. and I.C.; resources, B.B.K. and I.C.; data curation, M.R., I.C. and S.J.K.; writing—original draft preparation, E.I.A. and S.J.K.; writing—review and editing, M.R. and E.I.A.; visualization, E.I.A.; supervision, B.B.K.; project administration, B.B.K.; funding acquisition, S.J.K., K.K.S., S.B., R.K.S. and S.G.W. All authors have read and agreed to the published version of the manuscript.

Funding: This research received no external funding.

Institutional Review Board Statement: Not applicable.

Informed Consent Statement: Not applicable.

Data Availability Statement: Additional data will be made available on request.

Acknowledgments: Authors gratefully acknowledge the TÜBİTAK (The Scientific and Technological Research Council of Turkey), ULAKBİM (Turkish Academic Network and Information Centre), High Performance and Grid, Computing Centre (TRUBA resources), Turkey for providing necessary facilities for performing the computational studies. Authors extend their sincere thanks to Sagarika Chandra for her needful help towards editing the images incorporated in the manuscript.

Conflicts of Interest: The authors declare no conflict of interest.

References

1. Jamkhande, P.G.; Barde, S.R.; Patwekar, S.L.; Tidke, P.S. Plant profile, phytochemistry and pharmacology of *Cordia dichotoma* (Indian cherry): A review. *Asian Pac. J. Trop. Biomed.* **2013**, *3*, 1009–1012. [CrossRef]
2. Hussain, N.; Kakoti, B.B.; Rudrapal, M.; Junejo, J.A.; Laskar, M.A.; Lal, M.; Sarwa, K.K. Anticancer and Antioxidant Activities of *Cordia dichotoma* Forst. *Int. J. Green. Pharm.* **2020**, *14*, 265–273.
3. Hussain, N.; Kakoti, B.B.; Rudrapal, M.; Sarwa, K.K. Anti-inflammatory and Antioxidant Activities of *Cordia dichotoma* Forst. *Biomed. Pharmacol. J.* **2020**, *13*, 2093–2099. [CrossRef]
4. Hussain, N.; Kakoti, B.B.; Rudrapal, M.; Rahman, Z.; Rahman, M.; Chutia, D.; Sarwa, K.K. Antidiabetic Activity of the bark of Indian Cherry, *Cordia dichotoma*. *Biosci. Biotech. Res. Comm.* **2020**, *13*, 2211–2216. [CrossRef]
5. Ragasa, C.Y.; EbajoJr, V.D.; Mariquit, M.; Mandia, E.H.; Tan, M.C.S.; Brkljača, R.; Urban, S. Chemical constituents of *Cordia dichotoma* G. Forst. *J. Appl. Pharm. Sci.* **2015**, *5*, 16–21.
6. Junejo, J.A.; Rudrapal, M.; Mohammed, A.; Zaman, K. New flavonoid with antidiabetic potential from *Tetrastigma angustifolia* (Roxb.) Deb leaves. *Braz. J. Pharm. Sci.* **2020**, *56*, e18806. [CrossRef]
7. Junejo, J.A.; Mondal, P.; Rudrapal, M.; Zaman, K. Antidiabetic assessment of the hydro-alcoholicleaf extracts of the plant *Tetrastigma angustifolia* (Roxb.), a traditionally used North-Eastern Indian vegetable. *Biomed. Pharmacol. J.* **2014**, *7*, 635–644. [CrossRef]
8. Junejo, J.A.; Gogoi, G.; Islam, J.; Rudrapal, M.; Mondal, P.; Hazarika, H.; Zaman, K. Exploration ofantioxidant, antidiabetic and hepatoprotective activity of *Diplazium esculentum*, a wild edible plant from North Eastern region of India. *Future J. Pharm. Sci.* **2018**, *4*, 93–101. [CrossRef]
9. Kumar, S.; Kaushik, A.; Narasimhan, B.; Shah, S.A.; Lim, S.M.; Ramasamy, K.; Mani, V. Molecular docking, synthesis and biological significance of pyrimidine analogues as prospective antimicrobial and antiproliferative agents. *BMC Chem.* **2019**, *13*, 85. [CrossRef] [PubMed]
10. Wen, W.; Alseekh, S.; Fernie, A.R. Conservation and diversification of flavonoid metabolism in the plant kingdom. *Curr. Opin. Plant. Biol.* **2020**, *55*, 100–108. [CrossRef]
11. Arora, S.; Itankar, P. Extraction, isolation and identification of flavonoid from *Chenopodium album* aerial parts. *J. Trad. Comp. Med.* **2018**, *8*, 476–482. [CrossRef] [PubMed]
12. Kishore, N.; Twilley, D.; Blom van Staden, A.; Verma, P.; Singh, B.; Cardinali, G.; Lall, N. Isolation of flavonoids and flavonoid glycosides from *Myrsine africana* and their inhibitory activities against mushroom tyrosinase. *J. Nat. Prod.* **2018**, *81*, 49–56. [CrossRef] [PubMed]

13. Tundis, R.; Loizzo, M.R.; Menichini, F. Natural products as α-amylase and α-glucosidase inhibitors and their hypoglycaemic potential in the treatment of diabetes: An update. *Mini Rev. Med. Chem.* **2010**, *10*, 315–331. [CrossRef]

14. Junejo, J.A.; Zaman, K.; Rudrapal, M.; Hussain, N. Antidiabetic and Antioxidant Activity of Hydro-alcoholic Extract of Oxalis debilis Kunth. Leaves in Experimental Rats. *Biosci. Biotech. Res. Comm.* **2020**, *13*, 860–867. [CrossRef]

15. Junejo, J.A.; Rudrapal, M.; Zaman, K. Antidiabetic activity of *Carallia brachiata* Lour. Leaves hydro-alcoholic extract (HAE) with antioxidant potential in diabetic rats. *Indian J. Nat. Prod. Resour.* **2020**, *11*, 18–29.

16. Junejo, J.A.; Rudrapal, M.; Nainwal, L.M.; Zaman, K. Antidiabetic activity of hydro-alcoholic stem bark extract of *Callicarpa arborea* Roxb. with antioxidant potential in diabetic rats. *Biomed. Pharmacother.* **2017**, *95*, 84–94. [CrossRef]

17. Rashed, K.N.; Butnariu, M. Isolation and antimicrobial and antioxidant evaluation of bioactive compounds from *Eriobotrya japonica* stem. *Adv. Pharm. Bull.* **2014**, *4*, 75–81.

18. Conforti, F.; Statti, G.A.; Tundis, R.; Menichini, F.; Houghton, P. Antioxidant activity of methanolic extract of *Hypericum triquetrifolium* Turra aerial part. *Fitoterapia* **2002**, *73*, 479–483. [CrossRef]

19. Lynch, T.; Price, A.L. The effect of cytochrome P450 metabolism on drug response, interactions, and adverse effects. *Am. Fam. Physician.* **2007**, *76*, 391–396.

20. Feng, B.; Mills, J.B.; Davidson, R.E.; Mireles, R.J.; Janiszewski, J.S.; Troutman, M.D.; de Morais, S.M. In vitro P-glycoprotein assays to predict the in vivo interactions of P-glycoprotein with drugs in the central nervous system. *Drug Metab. Dispos.* **2008**, *36*, 268–275. [CrossRef]

21. Isyaku, Y.; Uzairu, A.; Uba, S. Computational studies of a series of 2-substituted phenyl-2-oxo-,2-hydroxyl-and 2-acylloxyethylsulfonamides as potent anti-fungal agents. *Heliyon* **2020**, *6*, e03724. [CrossRef]

22. Onah, O.E.; Babangida, K.J. Phytochemical Investigation and Antimicrobial Activity of Hexane, Ethyl Acetate and Methanol Fractions from Stem Bark of *Icacina trichantha* Oliv (Icacinaceae). *J. Chem. Environ. Sci. Appl.* **2020**, *7*, 7–12. [CrossRef]

23. Jigna, P.; Sumitra, C. Phytochemical screening of some plants from western region of India. *Plan. Arch.* **2008**, *8*, 657–662.

24. Roig-Zamboni, V.; Cobucci-Ponzano, B.; Iacono, R.; Ferrara, M.C.; Germany, S.; Bourne, Y.; Sulzenbacher, G. Structure of human lysosomal acid α-glucosidase–a guide for the treatment of Pompe disease. *Nat. Commun.* **2017**, *8*, 1–10. [CrossRef] [PubMed]

25. Nahoum, V.; Roux, G.; Anton, V.; Rougé, P.; Puigserver, A.; Bischoff, H.; Payan, F. Crystal structures of human pancreatic α-amylase in complex with carbohydrate and proteinaceous inhibitors. *Biochem. J.* **2000**, *346*, 201–208. [CrossRef] [PubMed]

26. Trott, O.; Olson, A.J. AutoDockVina: Improving the speed and accuracy of docking with a new scoring function, efficient optimization, and multi threading. *J. Comput. Chem.* **2010**, *31*, 455–461.

27. Daina, A.; Michielin, O.; Zoete, V. Swiss ADME: A free web tool to evaluate pharmacokinetics, drug-likeness and medicinal chemistry friendliness of small molecules. *Sci. Rep.* **2017**, *7*, 1–13. [CrossRef]

28. Kato-Schwartz, C.G.; deSá-Nakanishi, A.B.; Guidi, A.C.; Gonçalves, D.A.; Bueno, F.G.; Zani, B.P.M.; Peralta, R.M. Carbohydrate digestive enzymes are inhibited by *Poincianellapluviosa stem* bark extract: Relevance on type 2 diabetes treatment. *Clin. Phytosci.* **2020**, *6*, 1–11. [CrossRef]

 *molbank*

 MDPI

Article

2-Chloro-4,6-*bis*{(*E*)-3-methoxy-4-[(4-methoxybenzyl)oxy]styryl} pyrimidine: Synthesis, Spectroscopic and Computational Evaluation

Otávio Augusto Chaves [1], Vitor Sueth-Santiago [2,3,†], Douglas Chaves de Alcântara Pinto [2], José Carlos Netto-Ferreira [2,4], Debora Decote-Ricardo [5] and Marco Edilson Freire de Lima [2,*]

[1] Centro de Química de Coimbra (CQC), Departamento de Química, Universidade de Coimbra, Rua Larga, 3004-545 Coimbra, Portugal; otavioaugustochaves@gmail.com

[2] Departamento de Química Orgânica, Instituto de Química, Universidade Federal Rural do Rio de Janeiro, Seropédica 23890-000, Brazil; vitor_sueth@hotmail.co.uk (V.S.-S.); douglasdoti@hotmail.com (D.C.d.A.P.); jcnetto@ufrrj.br (J.C.N.-F.)

[3] Instituto Federal de Educação, Ciência e Tecnologia do Rio de Janeiro, Campus São Gonçalo, Rua José Augusto Pereira dos Santos, São Gonçalo 24425-004, Brazil

[4] Departamento de Química Orgânica, Instituto de Química, Universidade Federal do Rio de Janeiro, Centro de Tecnologia, Bloco A, Cidade Universitária, Rio de Janeiro 21941-909, Brazil

[5] Departamento de Microbiologia e Imunologia Veterinária, Instituto de Veterinária, Universidade Federal Rural do Rio de Janeiro, Seropédica 23890-000, Brazil; decotericardo@ufrrj.br

* Correspondence: marcoedilson@gmail.com

† This work is dedicated to the memory of the honorable Brazilian scientist, teacher, friend and human being Vitor Sueth-Santiago (1987–2021). A victim from COVID-19.

Citation: Chaves, O.A.; Sueth-Santiago, V.; Pinto, D.C.d.A.; Netto-Ferreira, J.C.; Decote-Ricardo, D.; Lima, M.E.F.d. 2-Chloro-4,6-*bis*{(*E*)-3-methoxy-4-[(4-methoxybenzyl)oxy]styryl}pyrimidine: Synthesis, Spectroscopic and Computational Evaluation. *Molbank* **2021**, *2021*, M1276. https://doi.org/10.3390/M1276

Academic Editor: Giovanni Ribaudo

Received: 31 July 2021
Accepted: 29 August 2021
Published: 7 September 2021

Publisher's Note: MDPI stays neutral with regard to jurisdictional claims in published maps and institutional affiliations.

Abstract: A novel curcumin analog namely 2-chloro-4,6-*bis*{(*E*)-3-methoxy-4-[(4-methoxybenzyl)oxy]styryl}pyrimidine (compound **7**) was synthesized by three-step reaction. The condensation reaction of protected vanillin with 2-chloro-4,6-dimethylpyrimidine (**6**) was the most efficient step, resulting in a total yield of 72%. The characterization of compound **7** was performed by ^{1}H and ^{13}C nuclear magnetic resonance (NMR), as well as high-resolution mass spectrometry. The experimental spectrometric data were compared with the theoretical spectra obtained by the density functional theory (DFT) method, showing a perfect match between them. UV-visible spectroscopy and steady-state fluorescence emission studies were performed for compound **7** in solvents of different polarities and the results were correlated with DFT calculations. Compound **7** showed a solvatochromism effect presenting higher molar extinction coefficient (log ε = 4.57) and fluorescence quantum yield (ϕ = 0.38) in toluene than in acetonitrile or methanol. The simulation of both frontier molecular orbitals (FMOs) and molecular electrostatic potential (MEP) suggested that the experimental spectra profile in toluene was not interfered by a possible charge transfer. These results are an indication of a low probability of compound **7** in reacting with unsaturated phospholipids in future applications as a fluorescent dye in biological systems.

Keywords: curcumin analog; organic synthesis; photophysical properties; steady-state fluorescence; DFT calculation

1. Introduction

Turmeric, obtained from the dried rhizomes of *Curcuma longa* (Zingiberaceae), is a golden colored material, commonly used around the world for seasoning and food coloring. Since antiquity, turmeric has been widely used in the treatment of several diseases in traditional Chinese and Indian medicine (traditionally known as Ayurveda, e.g., for the treatment of inflammatory diseases) [1–3]. Curcumin ((1*E*,6*E*)-1,7-*bis*(4-hydroxy-3-methoxyphenyl) hepta-1,6-diene-3,5-dione) also known as diferuloylmethane, is the main chemical component of turmeric (accounting for up to 70%), belonging to the class of diarylheptanoid metabolites. This compound is mainly responsible for both biological, metal chelator, flavoring, reactivity and pigment properties of turmeric (Figure 1A) [1,4]. Furthermore, curcumin stems present potent antioxidant, anti-inflammatory and anticancer activity due to their capacity to suppress the proliferation of a wide variety of tumor cells and regulate the expression of different enzymes [5]. Some structural modifications in the curcumin moiety based on structure-activity relationship have improved its drug profile (Figure 1). As an example, dimethoxycurcumin showed potential trypanocidal properties with a half-maximal inhibitory concentration (IC_{50}) value of 11.07 µM [6], while 1-methyl-3,5-*bis*[(*E*)-4-pyridyl) methylidene]-4-piperidone and 1-isopropyl-3,5-*bis*[(pyridine-3-yl) methylene] piperidin-4-one showed potential anticancer properties with IC_{50} values in the 0.7–1.0 µM range for H3122 cells lines (lung cancer cell lines) [7]. In addition, a series of asymmetric dihydrothiopyran curcumin analogs demonstrated high inhibition at a submicromolar level against acute promyelocytic leukemia cells [8], and integrating a tetrahydro-4-pyrone linker into curcumin structure led to cell growth inhibition, promoted apoptosis and enhanced irinotecan sensitivity against gastric cancer cells [9]. Finally, theoretical evaluation for some curcumin derivatives (replacing methoxyl groups in the aromatic moieties for bromo, chloro or hydroxyl groups) indicated these novel derivatives as potential inhibitors of amyloid-β peptides aggregation in the human brain (Aβs—aggregation process involved in the onset of Alzheimer's disease) [10].

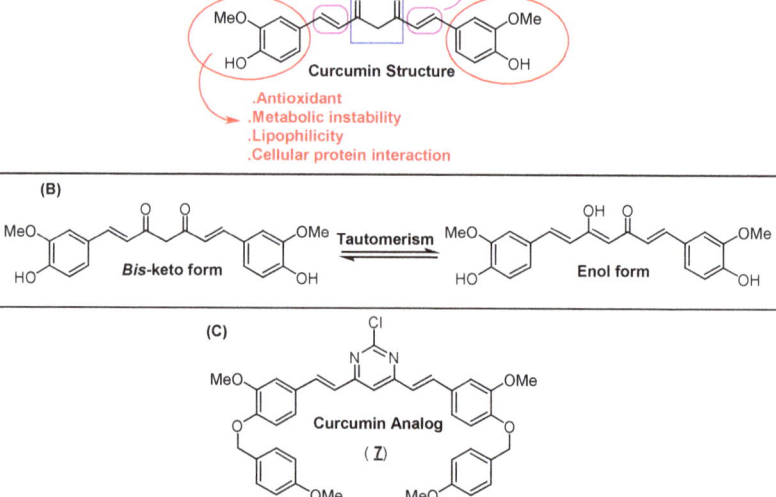

Figure 1. (**A**) Chemical structure of curcumin and its simplified structural-activity relationship. (**B**) Tautomerism form for curcumin. (**C**) Chemical structure for the curcumin analog under study—compound **7**.

From a structural point of view, the *ortho*-methoxy phenolic groups are symmetrically connected to a seven-carbon chain through an α,β-unsaturated β-diketone group. All these connections will result in a highly conjugated structure that shows UV-visible absorption in the 200–500 nm range, depending on the solvent polarity; however, the spectral properties of curcumin are related to the tautomeric forms (*bis*-keto or enol—Figure 1B) [11]. For this reason, curcumin has been widely explored in terms of spectroscopic properties, showing maximum absorption in the 408–430 nm range in different organic solvents; however, its maximum steady-state fluorescence emission (460–560 nm) is more sensitive than absorption to the solvent polarity, presenting Stokes' shift varying from 2000 to 6000 cm^{-1}.

Furthermore, curcumin shows a low fluorescence quantum yield in most of the solvents, which is significantly reduced in the presence of water [12]. In this sense, novel organic and inorganic compounds based on curcumin moiety have been proposed to improve the photophysical behavior, e.g., curcumin boron complexes can act as a near-infrared imaging fluorescent probe [13], chemical sensor [14] and larger second-order nonlinear property [15].

Based on the background described above on the biological and photophysical importance of curcumin and both its derivatives and analogs, the main goal of the present work is the synthesis of a novel symmetric curcumin analog, namely 2-chloro-4,6-*bis*{(*E*)-3-methoxy-4-[(4-methoxybenzyl)oxy]-styryl}pyrimidine (compound **7**—number according to the synthetic steps shown in Scheme 1) (Figure 1C). Its characterization via high-resolution mass spectrometry (HRMS) and experimental or computational Nuclear Magnetic Resonance spectra (NMR—^{1}H and ^{13}C) will follow the synthetic procedure. The preliminary spectroscopic characterization of compound **7** was also described by UV-visible and steady-state fluorescence techniques for three different solvents (methanol, acetonitrile or toluene), combined with computational results based on Density Functional Theory (DFT) calculations.

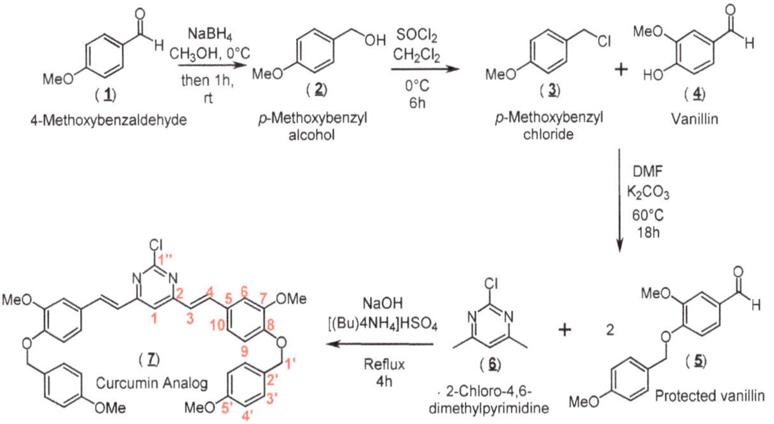

Scheme 1. Synthetic procedure to obtain the curcumin analog **7**.

2. Results and Discussion

2.1. Organic Synthesis and Structure Determination

The organic synthesis of compound **7** (curcumin analog) was performed through three main steps, as shown in Scheme 1. Basically, the first step was aimed at preparing para-methoxybenzyl chloride (**3**), which was employed in the derivatization of vanillin (**4**). Protected vanillin, namely 3-methoxy-4-[(4-methoxybenzyl)oxy]benzaldehyde (**5**), was then condensed with 2-chloro-4,6-dimethylpyrimidine (**6**) based on a protocol from Lee and coworkers [16], resulting in the formation of the target compound **7**.

Compound **7** was characterized by experimental and computational ^{1}H- and ^{13}C-NMR (Figures S1, S2 and Table S1 in the Supplementary Material), as well as by high resolution mass spectrometry (HRMS). As can be seen in Figure S1 and Table S1, the experimental ^{1}H-NMR signals (δ) at 2.50 and 3.30 correspond to DMSO-d$_6$ (solvent) and remnants hydration molecules, respectively, while δ at 3.77 and 3.85 correspond to the hydrogens for the methoxy groups connected to C$_{5'}$ and C$_7$, respectively. In addition, the signals (δ) in the 6.96–7.89 range were assigned to the hydrogens of the aromatic moiety of compound **7** (details in Figure S1). On the other hand, the experimental ^{13}C-NMR δ (via Distorsionless Enhancement by Polarization Transfer Including the Detection of Quaternary Nuclei—DEPTQ—Figure S2 and Table S1), revealed absorptions compatible with the proposed structure, e.g., δ at 55.58 and 56.09 corresponding to the carbon from the methoxyl groups connected to C$_{5'}$ and C$_7$, respectively. In addition, δ in the 111.00–166.28 range can be assigned to the carbons present in the aromatic moiety (details in Figure S2) of compound **7**. Overall, the proposed structure for **7** is in agreement with the experimental NMR data described in the literature for a similar compound: (E,E)-4,6-bis-(40-hydroxy-30-methoxystyryl)pyrimidine [16]. To further confirm the structure for compound **7**, HRMS experiments revealed the value of *m/z* 650.2184 [M]$^+$ which is in full agreement with the chemical formula $C_{38}H_{35}ClN_2O_6$ (calculated monoisotopic mass: 650.2184 g/mol).

Through the rise of reliable quantum chemical computational methods, such as Density Functional Theory (DFT), the DFT-NMR spectra of different organic compounds have been extensively reported. A comparison between the experimental and theoretical spectra in most cases reveals that they show the same profile [17–19]. Figures S1B and S2B show the theoretical DFT spectra (^{1}H- and ^{13}C-NMR) for the compound **7** and Table S1 compares the δ_{exp} and δ_{calc} values (experimental and theoretical, respectively). An inspection of the experimental and theoretical δ values revealed that they are practically the same in both cases. As an example, the methyl protons from methoxyl group in $C_{5'}$-OCH_3 presented δ_{calc} 3.77/δ_{exp} 3.67 and δ_{calc} 55.58/δ_{exp} 54.00 for ^{1}H- and ^{13}C-NMR, respectively. These results are clear indication that the chemical structure of the synthesized compound **7** is in accordance with the proposed structure. In addition, it was also shown that DFT is a good method to simulate NMR spectra for this curcumin analog.

2.2. Spectroscopic Study: UV-Vis Absorption and Steady-State Fluorescence Emission

In general, organic compounds based on curcumin are dyes whose structure allows them to be classified as donor-acceptor-donor species (D-A-D) due to the presence of electron donor groups at both ends of the conjugated π system and an electron acceptor group at their central portion [20]. The synthetic compound under study, i.e., compound **7**, also shows a D-A-D structure (Figure 2A), with a large absorption band in about 380, 390 and 395 nm in methanol, acetonitrile and toluene, respectively (Figure 2B and Table 1). These bands are probably due to a n-π* electronic transition, with an additional absorption at low wavelength (<325 nm), corresponding to a π-π* electronic transition in both methanol and acetonitrile [21]. It is important to note that toluene has a cut-off band of about 285 nm. Therefore, for toluene, only maximum absorption bands at 300 and 395 nm were considered. The absorption spectrum in each solvent (Figure 2B) shows a broad band in the high absorption wavelength region (325–450 nm range). The absence of shoulders in the absorption band in this region probably indicates that compound **7** does not present isomeric forms in the ground-state, which is in accordance with the proposed structure (Figure 1C). These results are in opposition to those described for curcumin, which show keto-enol tautomerism (Figure 1B) [22]. A considerable change in the energy of electronic transition in the 325–450 nm range can also be observed (Figure 2B) due to a solvatochromism effect, starting from a polar protic solvent (methanol) to a nonpolar solvent (toluene). For methanol, the presence of a blue shift in the maximum absorption band at 380 nm can be clearly observed, which is probably due to interactions by hydrogen bonding between the solvent and the electron acceptor groups (alkoxyl and the nitrogen atoms of the pyrimidine ring) present in compound **7** [12,22]. In addition, compound **7** showed a higher extinction coefficient (log ε at λ_{max} = 380, 390 and 395 nm for methanol, acetonitrile and toluene, respectively—Figure 2C and Table 1) in nonpolar and polar aprotic solvents (e.g., log ε = 4.57 in toluene), a spectroscopic behavior that is very suitable for photosensitizer and fluorescent dyes [12]. These results are similar to those reported in the literature for photosensitizers and fluorescent synthetic diacetoxyboron complexes, also based on curcumin structure [20].

Table 1. Photophysical results for the compound **7** in three different solvent polarities.

Solvent [a]	λ_{max} (nm)	Log ε [b]	λ_{exc} (nm)	λ_{em} (nm)	Φ_F	Stokes Shift (nm)	Dimer Formation
Methanol	220; 275; 380	4.19	222; 300; 392	507	0.09	115	No
Acetonitrile	225; 255; 390	4.48	225; 295; 395	504	0.23	109	No
Toluene	300; 395	4.57	304; 396	443	0.38	47	No

[a] UV-cutoff for methanol, acetonitrile and toluene: 210, 190 and 285 nm, respectively [23]. [b] Corresponding to the highest wavelength UV absorption (λ_{max} = 380, 390 and 395 nm for methanol, acetonitrile and toluene, respectively).

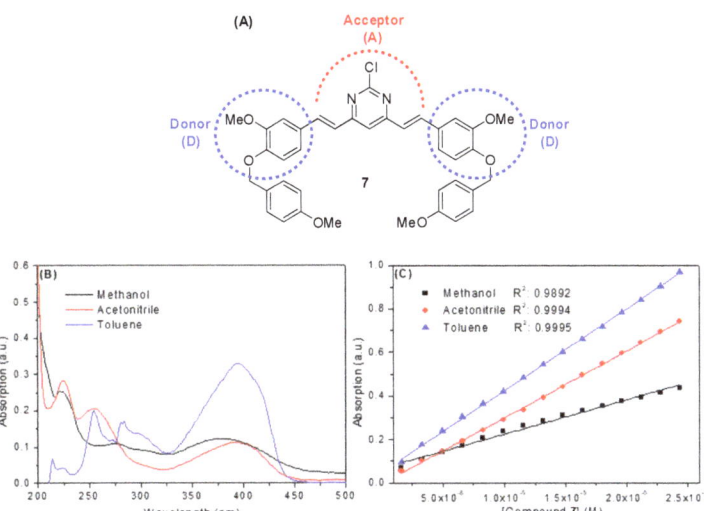

Figure 2. (A) Chemical structure of the compound **7** highlighting the D-A-D moieties. **(B)** UV-visible spectra for the compound **7** in methanol, acetonitrile and toluene. [**7**] = 2.32×10^{-6} mol/L **(C)** Beer-Lambert plots for the determination of ε value for **7** in three different solvents (at λ_{max} = 380, 390 and 395 for methanol, acetonitrile and toluene, respectively). [**7**] = 0.17–2.44×10^{-5} mol/L.

Steady-state fluorescence measurement is a comprehensive approach used to assess the excited state behavior of organic dyes in the presence of metallic species or solvents [20]. As can be seen in Figure 3A and Table 1, compound **7** shows maximum fluorescence emission in methanol, acetonitrile and toluene at 507, 504 and 443 nm, respectively (λ_{exc} = 390 nm, a wavelength in which none of these solvents contribute for the absorption phenomenon). By inspecting the spectra shown in Figure 3A, a significant red shift in the maximum fluorescence emission of compound **7** can be observed when the polarity of the solvent is varied. Thus, for toluene, a non-polar solvent, the maximum fluorescence emission for compound **7** was recorded at 443 nm. In contrast, for the polar solvents methanol or acetonitrile, maximum values were observed at 507 and 504 nm, respectively. The maximum fluorescence emission did not change significantly when using polar protic (methanol) or non-protic solvent (acetonitrile). Thus, it can be concluded that the process of solvation of the lowest energy unoccupied molecular orbital (LUMO) does not involve hydrogen bonding. Further confirmation of these results was made using computational calculations, which showed similar values, for both LUMO (−2.41 and −2.31 eV) and $|\Delta E|$ (3.29 and 3.27 eV) in methanol and acetonitrile, respectively (Table 2). These values justify the similarity between the spectroscopic data obtained in these two solvents. Surprisingly, compound **7** showed LUMO and $|\Delta E|$ values (−2.25 and 3.34 eV, respectively—Table 2) in toluene comparable in magnitude to those calculated for methanol and acetonitrile. However, the maximum fluorescence emission in the former solvent showed a significant blue-shift compared to that observed in the two later solvents (Figure 3).

Table 2. Comparison between experimental and theoretical (calculated—DFT) signals (δ) for ^{1}H- and ^{13}C-NMR to the compound **7**.

Solvent	μ (D)	E_{HOMO} (eV)	E_{LUMO} (eV)	$|\Delta E|$ (eV)
Methanol	6.26	−5.70	−2.41	3.29
Acetonitrile	6.21	−5.58	−2.31	3.27
Toluene	5.46	−5.59	−2.25	3.34

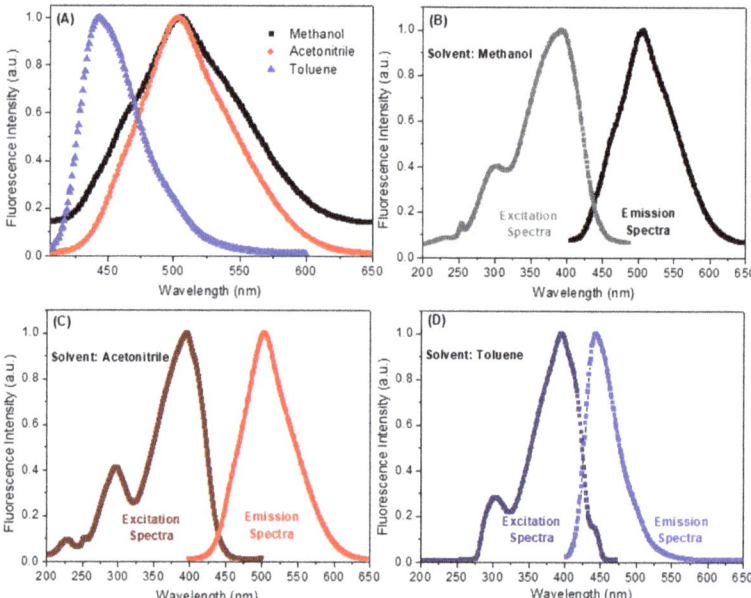

Figure 3. (**A**) Normalized steady-state fluorescence emission spectra for the compound **7** in three different solvents (λ_{exc} = 390 nm and [**7**] = 2.32 × 10^{-6} mol/L). (**B–D**) Normalized steady-state and excitation spectra for the compound **7** in methanol, acetonitrile and toluene, respectively.

The fluorescence quantum yield (Φ_F) for compound **7** was recorded in methanol, acetonitrile and toluene. All Φ_F determinations were performed using λ_{exc} = 390 nm for air saturated samples, with its value varying according to the polarity of the solvent. Compound **7** in methanol presented an extremely low fluorescence quantum yield (Φ_F = 0.09) when compared to acetonitrile (Φ_F = 0.23) and toluene (Φ_F = 0.38) (Table 1), suggesting that its singlet excited state in methanol must be deactivated mainly by a non-radiative process. The excitation spectrum for compound **7** (Figure 3B–D) showed a broadband in the 325–450 nm range, which can be attributed to the S$_0$–S$_1$ transition due to its fully superimposed to the corresponding UV–vis absorption spectrum (Figure 2) [22]. In addition, the excitation spectrum in all three solvents is the specular image of the corresponding fluorescence emission spectrum. The Stokes shift reflected the difference between the spectral position at the maximum for the excitation spectrum and the fluorescence emission, dependent on the fluorophore and the solvation environment. The maximum fluorescence emission for compound **7** in methanol, acetonitrile and toluene showed Stokes's shift of 115, 109 and 47 nm, respectively. These results agree with the computational ones, which indicated that more polar solvents typically lead to larger Stokes shifts [24].

In general, dye molecules with a large change in their permanent dipolar moment (μ) exhibit a strong solvatochromism after excitation [22,25]. As shown above, spectroscopic studies for compound **7**, regarding ground-state absorption and steady-state fluorescence emission indicated a different behavior depending on the solvent polarity. Thus, while the absorption spectrum for compound **7** showed a blue shift when the solvent polarity changed from non-polar to polar, a red shift was observed in the case of fluorescence emission spectrum (inverted solvatochromism). These results indicated that there must be a significant change in the dipole moment value (μ) for compound **7**, according to the polarity of the solvent [25]. To gain further insights into the behavior of the dipole moment for compound **7** in the presence of solvents of different polarities, the theoretical values for μ were calculated using the DFT method (Table 2). From these calculations, a significant change in the μ value was observed from methanol (6.26 D) to toluene (5.46 D). On the other hand, the μ value was quite similar in methanol and acetonitrile (6.26 to 6.21 D, respectively), also justifying the position of the maximum fluorescence in these two solvents, which was red-shifted when compared to toluene. The same solvatochromic profile for curcumin has already been reported [22]; however, these effects are mainly related to the keto-enol equilibrium, which is not the case for compound **7**, whose structure does not allow the existence of such equilibrium.

It is known that curcumin at a relatively high concentration can form dimers through a cycloaddition reaction due to its enolic form, in which a C=C double bond is present. This phenomenon directly affects the biological, photophysical and spectral profile of curcumin [26,27]. Even though the structure of compound **7** cannot lead to the formation of dimers due to the impossibility of the presence of an enolic form, it is important to theoretically assess its possible reactivity concerning the occurrence of a charge transfer process between compound **7** and model compounds that have an aromatic ring. The knowledge of this reactivity can help to understand the spectral profile discussed above when using toluene as a non-polar solvent.

The highest occupied molecular orbital (HOMO) and the lowest unoccupied molecular orbital (LUMO) are the most important frontier molecular orbitals (FMOs), playing a crucial role in understanding the stability and chemical reactivity of different compounds. In general, FMOs are responsible for predicting interactions between molecules (e.g., the interaction between small compounds and phospholipids in biological systems), electronic spectra and chemical reactions, such as the dimers' formation [28]. Theoretical studies on the curcumin reaction using FMOs indicated the contribution of HOMO and LUMO from the o-methoxyphenol and unsaturated β-diketone groups, respectively [26]. Figure 4A shows the theoretical FMOs representation (DFT method) for compound **7** in methanol, acetonitrile and toluene. The HOMO density for compound **7** is delocalized through the ortho-methoxyphenol portion and the α,β-unsaturated pyrimidine base. It was not found any evidence for the HOMO density delocalized over the oxygen connected to the aromatic group, which is probably due to the presence of the vanillin protected by the para-methoxybenzyl group. On the other hand, the LUMO density is located mainly in the α,β-unsaturated group linked to the pyrimidine ring [28]. Even though the LUMO density is in a possible reactive fraction of compound **7**, the fact that this region does not have reactive groups (as an example, a carbonyl group), in association with the presence of the protected vanillin, also affected the localized HOMO density. These results indicated that compound **7** has a low probability of reaction, including a charge transfer process in the presence of toluene.

An alternative that allows examining the reactive behavior of a molecule and can be used to understand the charge transfer process between a target compound and solvents, is the molecular electrostatic potential (MEP) map. The MEP is generated in space around a molecule by charge distribution, especially useful in understanding the sites for electrophilic attack and nucleophilic reactions in the study of biological recognition processes, hydrogen bonding interactions and dimer formation [28,29]. The prediction of the reactive molecular sites via MEP was obtained for compound **7** in the ground and excited singlet states (S_0 and S_1 states, respectively) by applying the DFT method in methanol, acetonitrile and toluene. Figure 4B shows the MEP maps with negative regions (assigned in red), corresponding to sites prone to electrophilic attack, and positive regions (assigned in blue) corresponding to nucleophilic reactivity. These computational results suggested that the possible sites for nucleophilic attack are found around the chlorine atom. However, this negative region is diluted in the α,β-unsaturated system both in the ground-state and excited singlet state, indicating a low possibility of reacting with other compounds, including charge transfer interaction, corroborating the FMOs' results. Overall, the computational results suggested low reactivity, which indicated that compound **7** probably would not react with unsaturated phospholipids in possible applications as a fluorescent dye in biological systems. Furthermore, these results also indicated a low probability of charge transfer between compound **7** and toluene, which cannot be responsible for the spectral profile for this compound in toluene [26]. In addition, the comparison between the MEP map and the Mulliken charge (Table 3) for the optimized S_0 and S_1 states revealed small changes in the distribution of the associated electronic charge in the ground and excited single states (only for the p-methoxyphenyl moiety). A slight increase was observed in the positive regions of the p-methoxyphenyl moiety; however, it is less rich in electrons when compared to the chlorine atom. These results suggested the pyrimidine ring contributes to the photophysical characteristics of compound **7** and the protected vanillin groups at its extremity can be responsible for possible further effects [30]. The fact that Mulliken's charge in the ground and excited states for compound **7** are practically the same when toluene is used as a solvent reinforced the low probability of charge transfer between this compound and non-polar aromatic solvents.

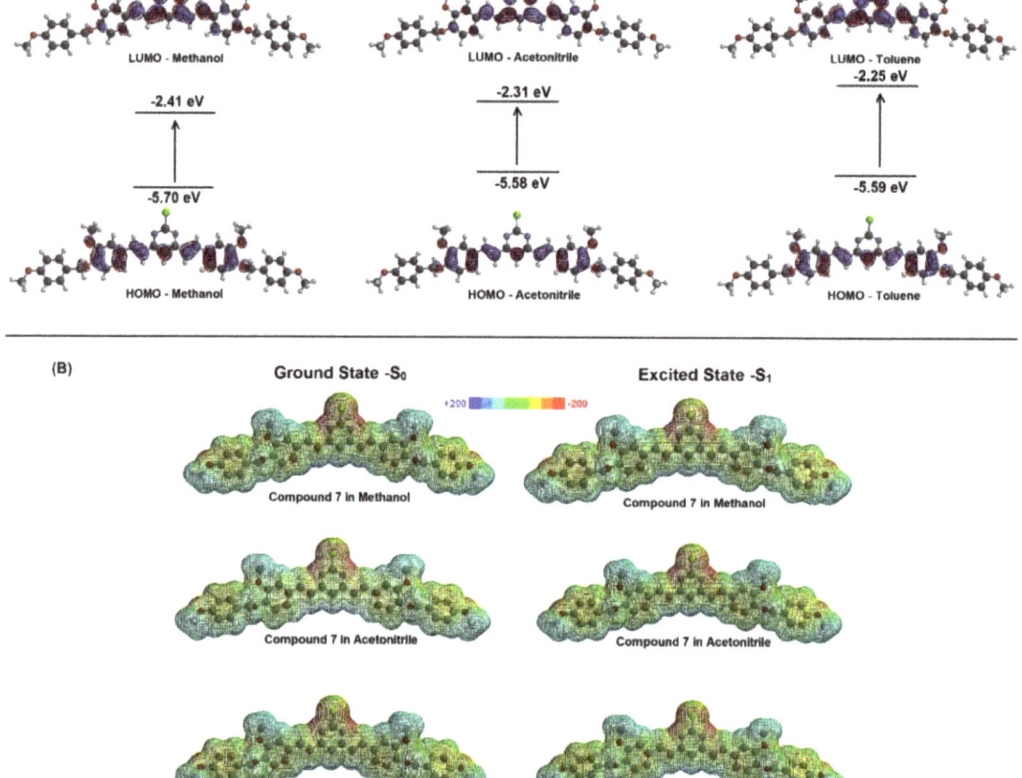

Figure 4. (**A**) HOMO-LUMO density for the compound **7** in methanol, acetonitrile and toluene. (**B**) Molecular electrostatic potential (MEP) map for the compound **7** (in a.u.) in three different solvents.

Table 3. Mulliken charge values for the compound **7** in both ground and excited state.

Position [a]	Mulliken Charge in the Ground State			Mulliken Charge in the Excited State		
	Methanol	Acetonitrile	Toluene	Methanol	Acetonitrile	Toluene
1	−0.194	−0.203	−0.218	−0.194	−0.203	−0.218
2	0.236	0.234	0.285	0.236	0.234	0.285
3	−0.185	−0.189	−0.192	−0.185	−0.189	−0.192
4	−0.193	−0.197	−0.185	−0.193	−0.197	−0.185
5	0.163	0.158	0.172	0.163	0.158	0.172
6	−0.270	−0.279	−0.281	−0.270	−0.279	−0.281
7	0.357	0.372	0.365	0.357	0.372	0.365
8	0.278	0.302	0.317	0.278	0.302	0.317
9	−0.189	−0.201	−0.190	−0.189	−0.201	−0.190
10	−0.184	−0.191	−0.193	−0.184	−0.191	−0.193
$1'$	−0.126	−0.116	−0.116	−0.126	−0.116	−0.116
$2'$	0.126	0.131	0.137	0.128	0.132	0.139
$3'$	−0.185	−0.182	−0.174	−0.182	−0.181	−0.174
$4'$	−0.198	−0.205	−0.190	−0.198	−0.205	−0.190
$5'$	0.351	0.376	0.380	0.351	0.376	0.382
$6'$	−0.198	−0.207	−0.190	−0.196	−0.206	−0.189
$7'$	−0.211	−0.220	−0.209	−0.210	−0.218	−0.207
$1''$	0.272	0.270	0.261	0.272	0.270	0.261
C_7-OCH$_3$	−0.242	−0.242	−0.232	−0.242	−0.242	−0.232
$C_{5'}$-OCH$_3$	−0.242	−0.240	−0.229	−0.242	−0.240	−0.229

[a] Carbon numeration according to denomination presented in Scheme 1.

3. Materials and Methods

All reagents and solvents were purchased from commercial sources (Tedia Ltda, Rio de Janeiro, Brazil and Sigma-Aldrich, Saint Louis, MO, USA) and used without further purification. The Nuclear Magnetic Resonance (NMR) spectra were recorded on a Bruker Ultrashield Plus Spectrometer (Bruker BioSpin GmbH, Rheinstetten, Germany) operating at 500 and 125 MHz for ^{1}H and ^{13}C (DEPTQ), respectively, with tetramethyl silane (TMS) as internal reference and deuterated dimethyl sulfoxide (DMSO-d$_6$) as solvent (signals for DMSO-d$_6$: δ 2.50 and 39.7 for ^{1}H and ^{13}C-NMR, respectively). Chemical shifts (δ) were reported in ppm and the coupling constants (*J*) in Hertz [Hz]. The High-Resolution Mass Spectrometry (HRMS) analyses were taken in the positive ion mode under electrospray ionization (ESI) method on a Bruker 9.4 T Apex-Qh (FT-ICR) (Bruker Daltonik GmbH Life Sciences, Bremen, Germany). Reactions were monitored by Thin Layer Chromatography (TLC) on Merck silica gel 60 F245 aluminum sheets and TLC spots were visualized by inspection of the plates under ultraviolet (UV) light (254 and 365 nm).

3.1. Synthesis of 2-Chloro-4,6-bis[(E)-3-methoxy-4-[(4-methoxybenzyl)oxy]-styryl]pyrimidine (Compound 7)

The organic synthesis of compound **7** was performed through three main steps: Firstly, synthesis of compound **3** was accomplished by reducing 1.80 mL of 4-methoxybenzaldehyde (**1**) (1.15 mmol) by sodium borohydride (NaBH$_4$). Initially, 0.28 g (7.5 mmol) of NaBH$_4$ was slowly added to a solution of (**1**) in 50 mL of methanol at 0 °C. The resulting solution (4-methoxybenzaldehyde + NaBH$_4$ in methanol) was left under stirring at room temperature for 1 h. The reaction was stopped by the addition of 50 mL of distilled water. The product was extracted with ethyl ether (3 × 50 mL) and the combined organic phases were washed with saturated sodium chloride solution (2 × 50 mL) and dried over anhydrous sodium sulfate (Na$_2$SO$_4$). The solvent was removed under reduced pressure without heating. A white solid (para-methoxybenzyl alcohol, **2**) was formed, which was then dissolved in 30 mL of dry dichloromethane (CH$_2$Cl$_2$) and cooled in an ice bath. Then, 1.5 mL of thionyl chloride (SOCl$_2$, 2.46 g, 20 mmol) was added dropwise to replace the hydroxyl group in **2** with a chlorine atom, a better leaving group for the second step of the reaction. The formed mixture was kept under a dry nitrogen atmosphere. Upon completion of the reaction, the solvent and excess SOCl$_2$ were removed by evaporation under reduced pressure. The crude product was isolated as a colorless oil with a quantitative yield and characterized by mass spectrometry (MS/MS: *m/z* 158 [M + 2]$^+$; 156 [M]$^+$; 121; 77; 51).

In the second synthetic step, in a bottom flask, compound **3** was solubilized in 30 mL of dry dimethylformamide (DMF), to which 2.20 g of vanillin (**4**, 15 mmol in DMF) and oven-dried potassium carbonate (K$_2$CO$_3$; 2.40 g, 18 mmol) were added. To remove the produced acid gas (HCl) without the entrance of atmospheric air, the flask containing the resulting suspension was coupled to a glass flowmeter and maintained at ambient pressure under stirring for 18 h. Then, the reaction mixture was poured into crushed ice, yielding a white crystalline solid corresponding to compound **5**, which was filtered and recrystallized from ethyl acetate. The yield (83%) was calculated from the initial amount of **1** used in the previous step, and compound **5** was characterized by mass spectrometry (MS/MS: *m/z* 272 [M]$^+$; 121; 77).

Finally, in the last step, 4 mmol of the protected vanillin (**5**) were condensed with 2 mmol of 2-chloro-4,6-dimethylpyrimidine (**6**) under 20 mL of an aqueous sodium hydroxide solution (NaOH, 4 mol/L) containing 0.29 mmol of tetrabutylammonium bisulfate ([(Bu)$_4$NH$_4$] HSO$_4$). This step was based on a protocol from Lee and coworkers [16]. The reaction mixture was refluxed until the formation of a yellow gum, which was then filtered and recrystallized from ethyl acetate, yielding a yellow solid (yield of 72% for the target compound **7**). ^{1}H-NMR (500 MHz, DMSO-d$_6$) δ (ppm): 7.87 (d, 2H, *J* = 16.0 Hz); 7.61 (s, 1H); 7.40 (m, 6H); 7.27 (d, 2H, *J* = 5.0 Hz); 7.20 (d, 2H, *J* = 16.0 Hz); 7.12 (d, 2H, *J* = 5.0 Hz); 6.96 (d, 4H, *J* = 7.0 Hz); 5.07 (s, 4H); 3.95 (s, 6H); and 3.77 (s, 6H). ^{13}C-NMR DEPTQ (125 MHz, DMSO-d$_6$) δ (ppm): 166.2 (C1''); 160.8 (C2); 159.5 (C5'); 150.1 (C8); 149.8 (C7); 138.7 (C4); 130.1 (C3'); 129.1 (C5); 128.6 (C2'); 123.1 (C3); 122.6 (C10); 114.9 (C6); 114.3 (C4'); 113.7 (C9); 111.0 (C1); 70.1 (C1'); 56.1 (C7-OCH$_3$); 55.6 (C5'-OCH$_3$). HRMS *m/z* 650.2184.

3.2. Spectroscopic Measurements

The UV-Vis spectra were measured in a Shimadzu model Mini 1240 (Shimadzu Scientific Instruments, Kyoto, Japan) at room temperature (ca 298 K). In a quartz cell (1.0 cm optical path), 3 mL of methanol, acetonitrile or toluene was used as solvent to measure the UV-spectra of compound **7** (concentration of 2.32 × 10^{-6} mol/L) in the 200–500 nm range. The molar extinction coefficient (ε) for compound **7** in the three different solvents (methanol, acetonitrile or toluene) was calculated using Beer-Lambert law (Equation (1)):

$$A = \varepsilon.l.c \tag{1}$$

where *l*, *A* and *c* are the cuvette cell pathlength, absorbance and molar concentration of the compound **7** (0.17–2.44 × 10^{-5} mol/L), respectively.

Steady-state fluorescence spectra measurements were performed using an optical spectrometer Jasco J-815 (Jasco Easton, MD, USA). A thermostated cuvette holder Jasco PFD-425S15F (Jasco Easton, MD, USA) was employed to control the temperature in the quartz cell (1.0 cm optical path). All spectra were recorded as the average of three scans with appropriate background corrections. The steady-state fluorescence spectrum of the compound **7** (concentration of 2.32 × 10^{-6} mol/L) in methanol, acetonitrile or toluene was obtained in the 406–650 nm range (λ_{exc} = 390 nm). The corresponding excitation spectrum was recorded at λ_{em} = 507, 504 and 443 nm in methanol, acetonitrile, and toluene, respectively. The evaluation of the aggregation process was carried out by recording the steady-state fluorescence spectrum upon successive additions of the compound **7** in the concentration range of 0.17–2.44 × 10^{-5} mol/L in all three different solvents (data shown only for acetonitrile). The fluorescence quantum yield (ϕ) at 298 K was calculated using Equation (2) and anthracene as a reference (λ_{exc} = 390 nm) [31]:

$$\varphi = \varphi_{ref} \cdot \frac{I_{sample}}{I_{ref}} \cdot \frac{A_{ref}^{390nm}}{A_{sample}^{390nm}} \cdot \frac{n^2}{n_{ref}^2} \tag{2}$$

where ϕ_{ref}, I, A^{390nm} and n^2 are the quantum yield for the reference compound (anthracene) [32], integral of the steady-state fluorescence emission spectra, absorbance at 390 nm (0.070 a.u.) and refractive index of the solvent, respectively. The subscript "ref" denotes the respective parameters for the reference compound (anthracene).

4. Conclusions

The curcumin analog 2-chloro-4,6-*bis*{(E)-3-methoxy-4-[(4-methoxybenzyl)oxy]-styryl} pyrimidine (compound **7**) was synthesized by three-step reaction. Basically, the first step was aimed at preparing *para*-methoxybenzyl chloride, which was employed in the derivatization of vanillin. Protected vanillin, namely 3-methoxy-4-((4-methoxybenzyl) oxy) benzaldehyde, was then condensed with 2-chloro-4,6-dimethylpyrimidine resulting in the formation of the target compound **7** with a final yield of 72%. The experimental and theoretical (DFT) signals of compound **7** by ^{1}H and ^{13}C-NMR confirmed the proposed structure, reinforced by HRMS *m/z* 650.2184 [M]$^{+}$ for $C_{38}H_{35}ClN_2O_6$. Compound **7** showed a large absorption band in about 380, 390 and 395 nm in methanol, acetonitrile and toluene, respectively, while steady-state fluorescence emission of compound **7** showed a solvatochromism effect with higher fluorescence quantum yield ($\phi = 0.38$) in toluene than in acetonitrile or methanol. The FMOs and MEP results indicated the absence of a possible charge transfer when toluene was used as solvent. Overall, the results indicated a low probability of compound **7** reacting with unsaturated phospholipids for future applications as a fluorescent dye in biological systems.

Supplementary Materials: The following are available online, Figures S1 and S2: The experimental and theoretical (DFT) ^{1}H- and ^{13}C-NMR spectra for the compound **7**. Table S1: Comparison between experimental and theoretical (calculated-DFT) signals (δ) for ^{1}H- and ^{13}C-NMR to the compound **7**.

Author Contributions: Conceptualization, M.E.F.d.L. and O.A.C.; synthesis and molecule characterization, V.S.-S., D.C.d.A.P. and M.E.F.d.L.; spectroscopic analysis and computational calculations, O.A.C. and J.C.N.-F.; writing—original draft preparation, O.A.C., V.S.-S. and M.E.F.d.L.; writing—review and editing, J.C.N.-F., M.E.F.d.L. and D.D.-R. All authors have read and agreed to the published version of the manuscript.

Funding: This research was funded by the Brazilian agencies: Coordenação de Aperfeiçoamento de Pessoal de Nível Superior (CAPES), Conselho Nacional de Desenvolvimento Científico e Tecnológico (CNPq) and Fundação Carlos Chagas Filho de Amparo à Pesquisa do Estado do Rio de Janeiro (FAPERJ). The funders had no role in the design of the study; in the collection, analyses or interpretation of data; in the writing of the manuscript or in the decision to publish the results.

Institutional Review Board Statement: Not applicable.

Informed Consent Statement: Not applicable.

Data Availability Statement: Not applicable.

Acknowledgments: The authors acknowledge Professor Nanci Câmara de Lucas Garden from Institute of Chemistry at Universidade Federal do Rio de Janeiro (UFRJ, Brazil) for the spectroscopic facilities. O.A.C. also thanks Fundação para a Ciência e a Tecnologia (FCT—Portuguese Foundation for Science and Technology) for the PhD fellowship 2020.07504.BD.

Conflicts of Interest: The authors declare no conflict of interest.

References

1. Sueth-Santiago, V.; Mendes-Silva, G.P.; Decoté-Ricardo, D.; de Lima, M.E.F. Curcumin, the golden powder from turmeric: Insights into chemical and biological activities. *Quím. Nova* **2015**, *38*, 538–552. [CrossRef]
2. Goel, A.; Kunnumakkara, A.B.; Aggarwal, B.B. Curcumin as "Curecumin": From kitchen to clinic. *Biochem. Pharm.* **2008**, *75*, 787–809. [CrossRef]
3. Khor, P.Y.; Aluwi, M.F.F.M.; Rullah, K.; Lam, K.W. Insights on the synthesis of asymmetric curcumin derivatives and their biological activities. *Eur. J. Med. Chem.* **2019**, *183*, 111704. [CrossRef] [PubMed]
4. Nelson, K.M.; Dahlin, J.L.; Bisson, J.; Graham, J.; Pauli, G.F.; Walters, M.A. The essential medicinal chemistry of curcumin. *J. Med. Chem.* **2017**, *60*, 1620–1637. [CrossRef] [PubMed]

5. Aggarwal, B.B.; Kumar, A.; Bharti, A.C. Anticancer potential of curcumin: Preclinical and clinical studies. *Anticancer Res.* **2003**, *23*, 363–398. [PubMed]

6. Sueth-Santiago, V.; Moraes, J.B.B.; Alves, E.S.S.; Vannier-Santos, M.A.; Freire-de-Lima, C.G.; Castro, R.N.; Mendes-Silva, G.P.; Del Cistia, C.N.; Magalhães, L.G.; Andricopulo, A.D.; et al. The effectiveness of natural diarylheptanoids against Trypanosoma cruzi: Cytotoxicity, ultrastructural alterations and molecular modeling studies. *PLoS ONE* **2016**, *11*, e0162926. [CrossRef]

7. Bland, A.R.; Bower, R.L.; Nimick, M.; Hawkins, B.C.; Rosengren, R.J.; Ashton, J.C. Cytotoxicity of curcumin derivatives in ALK positive non-small cell lung cancer. *Eur. J. Pharmacol.* **2019**, *865*, 172749. [CrossRef]

8. Tan, K.L.; Ali, A.; Du, Y.; Fu, H.; Jin, H.X.; Chin, T.M.; Khan, M.; Go, M.L. Synthesis and evaluation of bisbenzylidenedioxote-trahydrothiopranones as activators of endoplasmic reticulum (ER) stress signaling pathways and apoptotic cell death in acute promyelocytic leukemic cells. *J. Med. Chem.* **2014**, *57*, 5904–5918. [CrossRef] [PubMed]

9. Qiu, P.; Zhang, S.; Zhou, Y.; Zhu, M.; Kang, Y.; Chen, D.; Wang, J.; Zhou, P.; Li, W.; Xu, Q.; et al. Synthesis and evaluation of asymmetric curcuminoid analogs as potential anticancer agents that down regulate NF-kB activation and enhance the sensitivity of gastric cancer cell lines to irinotecan chemotherapy. *Eur. J. Med. Chem.* **2017**, *139*, 917–925. [CrossRef]

10. Shinzato, T.; Sato, R.; Suzuki, K.; Tomioka, S.; Sogawa, H.; Shulga, S.; Blume, Y.; Kurita, N. Proposal of therapeutic curcumin derivatives for Alzheimer's disease based on ab initio molecular simulations. *Chem. Phys. Lett.* **2002**, *738*, 136883. [CrossRef]

11. Waranyoupalin, R.; Wongnawa, S.; Wongnawa, M.; Pakawatchai, C.; Panichayupakaranant, P.; Sherdshoopongse, P. Studies on complex formation between curcumin and Hg(II) ion by spectrophotometric method: A new approach to overcome peak overlap. *Cent. Eur. J. Chem.* **2009**, *7*, 388–394. [CrossRef]

12. Priyadarsini, K.I. Photophysics, photochemistry and photobiology of curcumin: Studies from organic solutions, bio-mimetics and living cells. *J. Photochem. Photobiol. C* **2009**, *10*, 81–95. [CrossRef]

13. Ran, C.; Xu, X.; Raymond, S.B.; Ferrara, B.J.; Neal, K.; Bacskai, B.J.; Medarova, Z.; Moore, A. Design, synthesis, and testing of difluoroboron-derivatized curcumins as near infrared probes for in vivo detection of amyloid-β deposits. *J. Am. Chem. Soc.* **2009**, *131*, 15257–15261. [CrossRef]

14. Chaicham, A.; Kulchat, S.; Tumcharern, G.; Tuntulani, T.; Tomapatanaget, B. Synthesis, photophysical properties, and cyanide detection in aqueous solution of BF2-curcumin dyes. *Tetrahedron* **2010**, *66*, 6217–6223. [CrossRef]

15. Margar, S.N.; Rhyman, L.; Ramasami, P.; Sekar, N. Fluorescent difluoroboron-curcumin analogs: An investigation of the electronic structures and photophysical properties. *Spectrochim. Acta A* **2016**, *152*, 241–251. [CrossRef] [PubMed]

16. Lee, Y.S.; Kim, H.Y.; Kim, Y.S.; Seo, J.H.; Roh, E.J.; Han, H.; Shin, K.J. Small molecules that protect against β-amyloid-induced cytotoxicity by inhibiting aggregation of β-amyloid. *Bioorg. Med. Chem.* **2012**, *20*, 4921–4935. [CrossRef] [PubMed]

17. Da Silva, H.C.; De Almeida, W.B. Theoretical calculations of1H NMR chemical shifts for nitrogenated compounds in chloroform solution. *Chem. Phys.* **2020**, *528*, 110479. [CrossRef]

18. Soares, B.A.; Firme, C.L.; Maciel, M.A.M.; Kaiser, C.R.; Schilling, E.; Bortoluzzi, A.J. Experimental and NMR theoretical methodology applied to geometric analysis of the bioactive clerodane trans-dehydrocrotonin. *J. Braz. Chem. Soc.* **2014**, *25*, 629–638.

19. Souza, L.G.S.; Almeida, M.C.S.; Lemos, T.L.G.; Ribeiro, P.R.V.; Canuto, K.M.; Braz-Filho, R.; Del Cistia, C.N.; Sant'Anna, C.M.R.; Barreto, F.S.; de Moraes, M.O. Brazoides A-D, New Alkaloids from *Justicia gendarussa Burm.* F. Species. *J. Braz. Chem. Soc.* **2017**, *28*, 1281–1287. [CrossRef]

20. Lyu, H.; Wang, D.; Cai, L.; Wang, D.-J.; Li, X.-M. Synthesis, photophysical and solvatochromic properties of diacetoxyboron complexes with curcumin derivatives. *Spectrochim. Acta A* **2019**, *220*, 117126. [CrossRef]

21. Balasubramanian, K. Theoretical calculations on the transition energies of the UV-visible spectra of curcumin pigment in turmeric. *Ind. J. Chem. A* **1991**, *30*, 61–65.

22. Patra, D.; Barakat, C. Synchronous fluorescence spectroscopic study of solvatochromic curcumin dye. *Spectrochim. Acta A* **2011**, *70*, 1034–1041. [CrossRef]

23. Przybytek, J.T. *High-Purity Solvent Guide*, 1st ed.; Burdick & Jackson Laboratories: Muskegon, MI, USA, 1980.

24. Lakowicz, J.R. ; *Principles of Fluorescence Spectroscopy*, 3rd ed.; Springer: New York, NY, USA, 2006.

25. Jacques, P. On the relative contributions of nonspecific and specific interactions to the unusual solvatochromism of a typical merocyanine dye. *J. Phys. Chem.* **1986**, *90*, 5535–5539. [CrossRef]

26. Fujisawa, S.; Atsumi, T.; Ishihara, M.; Kadoma, Y. Cytotoxicity, ROS-generation activity and radical-scavenging activity of curcumin and related compounds. *Anticancer Res.* **2004**, *24*, 563–570.

27. Masuda, T.; Toi, Y.; Bando, H.; Maekawa, T.; Takeda, Y.; Yamaguchi, H. Structural identification of new curcumin dimers and their contribution to the antioxidant mechanism of curcumin. *J. Agric. Food Chem.* **2002**, *50*, 2524–2530. [CrossRef] [PubMed]

28. Kosar, B.; Albayrak, C. Spectroscopic investigations and quantum chemical computational study of (E)-4-methoxy-2-[(p-tolylimino)methyl]phenol. *Spectrochim. Acta A* **2011**, *78*, 160–167. [CrossRef]

29. Politzer, P.; Lane, P. A computational study of some nitrofluoromethanes. *Struct. Chem.* **1990**, *1*, 159–164. [CrossRef]

30. Lisboa, C.S.; de Lucas, N.C.; Garden, S.J. Synthesis of novel substituted methoxybenzo[2,3-b]carbazole derivatives via C-H functionalization. Experimental and theoretical characterization of their photophysical properties. *Dyes Pig.* **2016**, *134*, 618–632. [CrossRef]

31. Coppo, R.L.; Zanoni, K.P.S.; Iha, N.Y.M. Unraveling the luminescence of new heteroleptic Ir(III) cyclometalated series. *Polyhedron* **2019**, *163*, 161–170. [CrossRef]

32. Katoh, R.; Suzuki, K.; Furube, A.; Kotani, M.; Tokumaru, K. Fluorescence quantum yield of aromatic hydrocarbon crystals. *J. Phys. Chem. C* **2009**, *113*, 2961–2965. [CrossRef]

MDPI

Short Note

2-Oxo-2H-chromen-7-yl 4-chlorobenzoate

Diana Becerra [1,*], Jaime Portilla [2] and Juan-Carlos Castillo [1,2,*]

[1] Escuela de Ciencias Química, Facultad de Ciencias, Universidad Pedagógica y Tecnológica de Colombia, Avenida Central del Norte 39-115, Tunja 150003, Colombia

[2] Bioorganic Compounds Research Group, Department of Chemistry, Universidad de los Andes, Carrera 1 No. 18A-10, Bogotá 111711, Colombia; jportill@uniandes.edu.co

* Correspondence: diana.becerra08@uptc.edu.co (D.B.); juan.castillo06@uptc.edu.co (J.-C.C.); Tel.: +57-8740-5626 (ext. 2425) (D.B. & J.-C.C.)

Abstract: We describe the synthesis of 2-oxo-2H-chromen-7-yl 4-chlorobenzoate **3** in 88% yield by the O-acylation reaction of 7-hydroxy-2H-chromen-2-one **1** with 4-chlorobenzoyl chloride **2** in dichloromethane using a slight excess of triethylamine at 20 °C for 1 h. The ester **3** was completely characterized by mass spectrometry, IR, UV–Vis, 1D, and 2D NMR spectroscopy.

Keywords: 7-hydroxy-2H-chromen-2-one; O-acylation reaction; coumarin

Citation: Becerra, D.; Portilla, J.; Castillo, J.-C. 2-Oxo-2H-chromen-7-yl 4-chlorobenzoate. *Molbank* **2021**, *2021*, M1279. https://doi.org/10.3390/M1279

Academic Editor: Giovanni Ribaudo

Received: 29 August 2021
Accepted: 10 September 2021
Published: 14 September 2021

Publisher's Note: MDPI stays neutral with regard to jurisdictional claims in published maps and institutional affiliations.

1. Introduction

The coumarin was first isolated from tonka beans by A. Vogel in 1820 [1], while W. H. Perkin described the first chemical synthesis in 1868 by heating acetic acid with the sodium salt of salicylaldehyde [2]. The coumarin is also known as 2H-chromen-2-one (1,2-benzopyrone or 2H-1-benzopyran-2-one) according to the IUPAC nomenclature. This oxa-heterocycle is a two-ring system, consisting of a benzene ring fused with a α-pyrone nucleus. It should be noted that coumarin-based fluorescent chemosensors have been widely employed in bioorganic chemistry, molecular recognition, and materials science [3]. Over the last decades, synthetic and naturally occurring coumarins have received considerable attention from organic and medicinal chemists due to their huge diversity of biological and pharmacological activities, including anti-inflammatory [4], antibacterial [5], antifungal [6], anticoagulant [7], antioxidant [8], antiviral [9], cholinesterase (ChE), and monoamine oxidase (MAO) inhibitory properties [10]. Besides, coumarins exhibited significant anticancer activity through diverse mechanisms of action, including inhibition of carbonic anhydrase, inhibition of microtubule polymerization, inhibition of tumor angiogenesis, regulating the reactive oxygen species, among others [11–14].

In particular, 7-hydroxycoumarin derivatives have been widely used as valuable building blocks for the preparation of novel coumarin-based anticancer agents [15–17]. For instance, umbelliferone analogs (**I**) and (**II**) had excellent activity against MCF-7 cells with IC_{50} values of 9.54 and 16.1 μM, respectively, as illustrated in Figure 1 [15]. Interestingly, the coumarin-containing ketone (**III**) showed potent activity against breast cancer MCF-7 cells, with an IC_{50} value of 0.47 μM [16]. In contrast, the coumarin-containing ester (**IV**) exhibited high selectivity towards tumor-associated hCA IX over the cytosolic hCA I isoform, with a value of 21.8 nM [17].

It should be noted that the post-functionalization of the 7-hydroxycoumarin skeleton has been scarcely studied in synthetic and medicinal chemistry [18]. Interestingly, the hydroxyl group at the 7-position of the coumarin skeleton can be exploited to perform alkylation and acylation reactions [19–22]. Herein, we describe the synthesis and complete characterization of 2-oxo-2H-chromen-7-yl 4-chlorobenzoate **3** through an O-acylation reaction of 7-hydroxy-2H-chromen-2-one **1** with 4-chlorobenzoyl chloride **2** in the presence of triethylamine under mild reaction conditions.

61

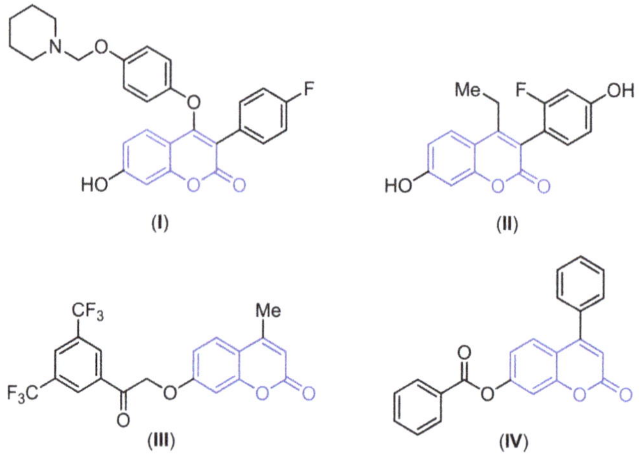

Figure 1. Biologically active 7-hydroxycoumarin derivatives.

2. Results and Discussion

In connection with the ongoing development of efficient and simple protocols for the acylation of heterocyclic compounds of biological interest [23,24], we describe an expeditious approach to synthesize 2-oxo-2*H*-chromen-7-yl 4-chlorobenzoate **3** through an *O*-acylation reaction between equimolar amounts of 7-hydroxy-2*H*-chromen-2-one **1** and 4-chlorobenzoyl chloride **2** in dichloromethane, using a slight excess of triethylamine with vigorous stirring at 20 °C for 1 h under normal atmospheric conditions (Scheme 1). After the specified reaction time, the solvent was removed under vacuum using a rotary evaporator. The resulting crude product was purified by flash chromatography on silica gel using dichloromethane as an eluent to furnish ester **3** in 88% yield. This procedure is distinguished by its short reaction times, high yield, clean reaction profile, and operational simplicity. Albeit the compound **3** was synthesized nine years ago [19], the structural and electronic information obtained from spectroscopic and spectrometry data has not been explained yet. For that reason, a complete spectroscopic and analytical characterization was performed in this work (see Section 3). Initially, the structure of **3** was determined by mass spectrometry, IR, UV–Vis, and 1D NMR spectroscopy (Figures S1–S7). Later, the analysis of 2D NMR spectra, including HSQC (Figure S8), HMBC (Figures S9 and S10), COSY (Figure S11), and NOESY (Figure S12), allowed the structural assignment without ambiguity.

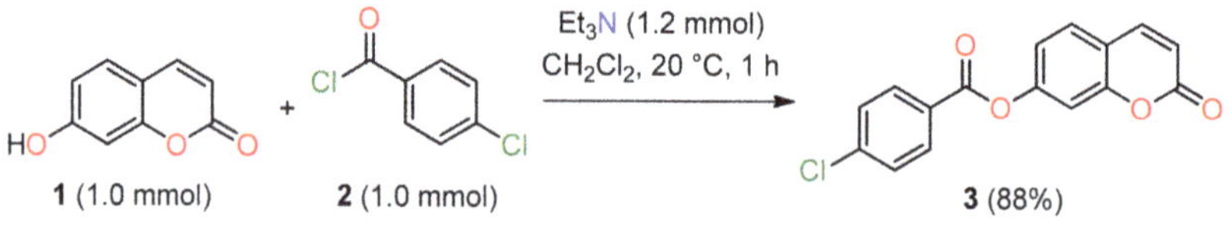

Scheme 1. Time-efficient synthesis of 2-oxo-2*H*-chromen-7-yl 4-chlorobenzoate **3**.

The absorption bands at 1728 and 1589/1620 cm^{-1} are assigned to the C=O and C=C stretching vibrations in the IR spectrum, respectively. The absorption bands at 1068/1092 and 1231/1261 cm^{-1} are attributed to the C–O–C asymmetric stretching vibrations. It should be noted that the C–Cl stretching band is normally expected around 580–750 cm^{-1} [25]; thus, a strong band at 744 cm^{-1} is assigned to the C–Cl stretching vibration. The ^{1}H-NMR spectrum of **3** recorded in DMSO-d_6 showed one doublet of doublets at 7.34 ppm and four doublets at 6.51, 7.48, 7.83, and 8.11 ppm for the coumarin ring, as well as two doublets at 7.70 and 8.15 ppm for the benzene ring (Table 1). The proton signal of the hydroxyl group attached to the coumarin ring was not observed, indicating that the *O*-acylation process was successful. The $^{13}C\{^1H\}$ NMR and DEPT spectra of **3** showed 14 carbon signals, consisting of seven aromatic methines, five quaternary aromatic carbons, and two carbonyl carbons (Table 1 and Figure 2A). The complete assignment of the proton and carbon signals of **3** is described in Section 3, while the correlations ^1H-^1H and ^1H-^{13}C observed in COSY and HMBC experiments, respectively, are illustrated in Figure 2B. In the MS spectrum, two molecular peaks are observed at m/z 300 and 302 complying with the Cl-rule, along with two peaks at m/z 141 and 139 with 32% and 100% intensity, respectively, corresponding to the (4-chlorobenzylidyne)oxonium ion ($C_7H_4ClO^+$). Additionally, the accurate mass (m/z 301.0261) of the pseudo-molecular ion ($[M + H]^+$) and the elemental formula ($C_{16}H_{10}ClO_4^+$) is confirmed by HRMS measurements, obtaining an error mass of 1.33 ppm.

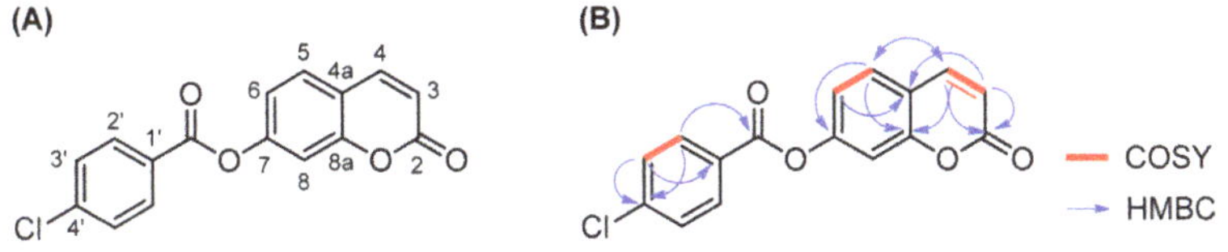

(A)   **(B)**

Figure 2. (**A**) Structure of 2-oxo-2*H*-chromen-7-yl 4-chlorobenzoate **3**. (**B**) Connectivities of **3** based on COSY (bold red line) and HMBC (from H to C, blue arrow) data.

Table 1. ^1H and $^{13}C\{^1H\}$ NMR assignments, and COSY, NOESY, and HMBC correlations of **3** [a].

Number	δ_H (mult, *J* in Hz)	δ_C (ppm)	COSY (^1H-^1H)	NOESY (^1H-^1H)	HMBC (^1H-^{13}C)
2	–	159.7	–	–	H-3 (^2J) H-4 (^3J)
3	6.51 (d, *J* = 9.6)	115.8	H-4 (^3J)	H-4	–
4	8.11 (d, *J* = 9.6)	143.9	H-3 (^3J)	H-3 H-5	H-5 (^3J)
4a	–	117.0	–	–	H-3 (^3J) H-6 (^3J)
5	7.83 (d, *J* = 8.4)	129.5	H-6 (^3J)	H-4 H-6	H-4 (^3J)
6	7.34 (dd, *J* = 8.4, 2.0)	118.8	H-5 (^3J)	H-5	–
7	–	152.9	–	–	H-5 (^3J)
8	7.48 (d, *J* = 2.0)	110.4	–	–	–
8a	–	154.1	–	–	H-4 (^3J) H-5 (^3J)
1'	–	127.4	–	–	H-3' (^3J)
2'	8.15 (d, *J* = 8.4)	131.8	H-3' (^3J)	H-3'	–
3'	7.70 (d, *J* = 8.4)	129.2	H-2' (^3J)	H-2'	–
4'	–	139.3	–	–	H-2' (^3J) H-3' (^2J)
C=O	–	163.4	–	–	H-2' (^3J)

[a] Measured at 400 MHz (^1H) and 101 MHz (^{13}C) in DMSO-d_6 at 25 °C.

In summary, we described the expeditious and ambient-temperature synthesis of 2-oxo-2*H*-chromen-7-yl 4-chlorobenzoate **3** through an *O*-acylation reaction of 7-hydroxy-2*H*-chromen-2-one **1** with 4-chlorobenzoyl chloride **2** in dichloromethane, using a slight excess of triethylamine. This protocol is distinguished by its short reaction times, high yield, clean reaction profile, and operational simplicity.

3. Materials and Methods

3.1. General Information

The 7-hydroxy-2*H*-chromen-2-one **1** (CAS 93-35-6) and 4-chlorobenzoyl chloride **2** (CAS 122-01-0) were purchased from Sigma–Aldrich (Saint Louis, MO, USA). The starting materials were weighed and handled in air at ambient temperature. The silica gel aluminum plates (Merck 60 F_{254}, Darmstadt, Germany) were used for analytical TLC. The IR absorption spectrum was recorded at room temperature employing a Shimadzu FTIR 8400 spectrophotometer (Scientific Instruments Inc., Seattle, WA, USA) equipped with an attenuated reflectance accessory. ^{1}H and ^{13}C{^{1}H} NMR spectra were recorded at 25 °C on a Bruker Avance 400 spectrophotometer (Bruker BioSpin GmbH, Rheinstetten, Germany) operating at 400 MHz and 101 MHz, respectively. The concentration of the sample was approximately 15 mg/0.5 mL of DMSO-d_6. Chemical shifts of ^{1}H and ^{13}C{^{1}H} NMR experiments were referenced by tetramethylsilane (δ = 0.0 ppm). Chemical shifts (δ) are given in ppm and coupling constants (J) are given in Hz. The 2D HSQC, HMBC, COSY, and NOESY experiments were performed using the standard Bruker pulse sequence. NMR data were analyzed using the MestReNova 12.0.0 (2017) software (Mestrelab, Escondido, CA, USA). The mass spectrum was recorded on a SHIMADZU-GCMS 2010-DI-2010 spectrometer (Scientific Instruments Inc., Columbia, WA, USA) equipped with a direct inlet probe operating at 70 eV. The high resolution mass spectrum (HRMS) was recorded using a Q-TOF spectrometer via electrospray ionization (ESI, 4000 V). The UV–Vis spectrum was obtained from an acetone solution (5.0×10^{-4} M) in an Evolution 201 UV–Vis spectrophotometer (Thermo Fischer Scientific Inc., Madison, WI, USA).

3.2. Synthesis of 2-Oxo-2H-Chromen-7-yl 4-Chlorobenzoate 3

A mixture of 7-hydroxy-2*H*-chromen-2-one **1** (162 mg, 1.0 mmol), 4-chlorobenzoyl chloride **2** (128 μL, 1.0 mmol), and triethylamine (167 μL, 1.2 mmol) in dichloromethane (5.0 mL) was stirred at 20 °C for 1 h (Scheme 1). After a complete disappearance of the starting materials, as monitored by thin-layer chromatography (TLC), the solvent was removed using a rotary evaporator under vacuum. The resulting crude product was purified by flash chromatography on silica gel using dichloromethane as an eluent to afford 2-oxo-2*H*-chromen-7-yl 4-chlorobenzoate **3** as colorless, needle-like crystals (265 mg, 88% yield): Rf (DCM) = 0.38. M.p 228–229 °C. FTIR-ATR: ν = 3086, 1728 (ν C=O), (1620 and 1589 for ν C=C), 1497, 1396, (1261 and 1231 for ν_a C–O–C), (1092 and 1068 for ν_a C–O–C), 984, (880 and 837 for ν_s C–O–C), 744 (ν C–Cl), 613, 540, 521 cm^{-1}. UV–Vis (acetone) λ_{max} (ε, L·mol^{-1}·cm^{-1}): 316 (469), 330 (3898) nm. ^{1}H-NMR (400 MHz, DMSO-d_6): δ = 6.51 (d, J = 9.6 Hz, 1H, H-3), 7.34 (dd, J = 8.4, 2.0 Hz, 1H, H-6), 7.48 (d, J = 2.0 Hz, 1H, H-8), 7.70 (d, J = 8.4 Hz, 2H, H-3′), 7.83 (d, J = 8.4 Hz, 1H, H-5), 8.11 (d, J = 9.6 Hz, 1H, H-4), 8.15 (d, J = 8.4 Hz, 2H, H-2′) ppm. ^{13}C{^{1}H}-NMR (101 MHz, DMSO-d_6): δ = 110.4 (CH, C-8), 115.8 (CH, C-3), 117.0 (Cq, C-4a), 118.8 (CH, C-6), 127.4 (Cq, C-1′), 129.2 (2CH, C-3′), 129.5 (CH, C-5), 131.8 (2CH, C-2′), 139.3 (Cq, C-4′), 143.9 (CH, C-4), 152.9 (Cq, C-7), 154.1 (Cq, C-8a), 159.7 (Cq, C-2), 163.4 (Cq, C=O) ppm. MS (EI, 70 eV) *m/z* (%): 302/300 (3/8) [M$^{+\bullet}$], 141/139 (32/100), 113/111 (28/85), 105 (14), 75 (29), 51 (16). HRMS (ESI+): calcd for $C_{16}H_{10}ClO_4^+$, 301.0257 [M + H]$^+$; found, 301.0261.

Supplementary Materials: The following are available online. Figure S1: HRMS spectrum for compound **3**; Figure S2: EIMS spectrum of the compound **3**; Figure S3: IR spectrum for compound **3**; Figure S4: UV–Vis spectrum for compound **3**; Figure S5: ^{1}H-NMR spectrum for compound **3**; Figure S6: ^{13}C{^{1}H} NMR and DEPT-135 spectra for compound **3**; Figure S7: Expansion ^{13}C{^{1}H} NMR and DEPT-135 spectra for compound **3**; Figure S8: HSQC 2D C–H correlation spectrum for compound **3**; Figure S9: HMBC 2D C–H correlation spectrum for compound **3**; Figure S10: Expansion HMBC 2D C–H correlation spectrum for compound **3**; Figure S11: COSY 2D H–H correlation spectrum for compound **3**; Figure S12: NOESY 2D H–H correlation spectrum for compound **3**.

Author Contributions: Investigation, data curation, writing—original draft preparation, D.B.; writing—review and editing, resources, J.P.; conceptualization, data curation, writing—original draft preparation, J.-C.C. All authors have read and agreed to the published version of the manuscript.

Funding: The APC was sponsored by MDPI.

Institutional Review Board Statement: Not applicable.

Informed Consent Statement: Not applicable.

Data Availability Statement: The data presented in this study are available in this article.

Acknowledgments: The authors thank Universidad Pedagógica y Tecnológica de Colombia and Universidad de los Andes. D.B. and J.-C.C. acknowledge to the Dirección de Investigaciones at the Universidad Pedagógica y Tecnológica de Colombia (Project SGI-3073). J.P. thanks support from the Facultad de Ciencias at the Universidad de los Andes (Project INV-2019-84-1800).

Conflicts of Interest: The authors declare not conflict of interest.

References

1. Vogel, A. Darftellung von benzoefäure aus der tonka-bohns und aus den meliloten-oder steinklee-blumen. *Ann. Phys.* **1820**, *64*, 161–166. [CrossRef]
2. Perkin, W.H. VI.—On the artificial production of coumarin and formation of its homologues. *J. Chem. Soc.* **1868**, *21*, 53–63. [CrossRef]
3. Cao, D.; Liu, Z.; Verwilst, P.; Koo, S.; Jangjili, P.; Kim, J.S.; Lin, W. Coumarin-based small-molecule fluorescent chemosensors. *Chem. Rev.* **2019**, *119*, 10403–10519. [CrossRef] [PubMed]
4. Grover, J.; Jachak, S.M. Coumarins as privileged scaffold for anti-inflammatory drug development. *RSC Adv.* **2015**, *5*, 38892–38905. [CrossRef]
5. Insuasty, D.; Castillo, J.; Becerra, D.; Rojas, H.; Abonia, R. Synthesis of biologically active molecules through multicomponent reactions. *Molecules* **2020**, *25*, 505. [CrossRef]
6. Zhang, S.; Tan, X.; Liang, C.; Zhang, W. Design, synthesis, and antifungal evaluation of novel coumarin-pyrrole hybrids. *J. Heterocycl. Chem.* **2021**, *58*, 450–458. [CrossRef]
7. Venugopala, K.N.; Rashmi, V.; Odhav, B. Review on natural coumarin lead compounds for their pharmacological activity. *Biomed. Res. Int.* **2013**, *2013*, 963248. [CrossRef]
8. Katsori, A.-M.; Hadjipavlou-Litina, D. Coumarin derivatives: An updated patent review (2012–2014). *Expert Opin. Ther. Patents* **2014**, *24*, 1323–1347. [CrossRef]
9. Hassan, M.Z.; Osman, H.; Ali, M.A.; Ahsan, M.J. Therapeutic potential of coumarins as antiviral agents. *Eur. J. Med. Chem.* **2016**, *123*, 236–255. [CrossRef]
10. Stefanachi, A.; Leonetti, F.; Pisani, L.; Catto, M.; Carotti, A. Coumarin: A natural, privileged and versatile scaffold for bioactive compounds. *Molecules* **2018**, *23*, 250. [CrossRef]
11. Musa, A.M.; Cooperwood, J.S.; Khan, M.O.F. A review of coumarin derivatives in pharmacotherapy of breast cancer. *Curr. Med. Chem.* **2008**, *15*, 2664–2679. [CrossRef]
12. Thakur, A.; Singla, R.; Jaitak, V. Coumarins as anticancer agents: A review on synthetic strategies, mechanism of action and SAR studies. *Eur. J. Med. Chem.* **2015**, *101*, 476–495. [CrossRef]
13. Zhang, L.; Xu, Z. Coumarin-containing hybrids and their anticancer activities. *Eur. J. Med. Chem.* **2019**, *181*, 111587. [CrossRef]
14. Wu, Y.; Xu, J.; Liu, Y.; Zeng, Y.; Wu, G. A review on anti-tumor mechanisms of coumarins. *Front. Oncol.* **2020**, *10*, 592853. [CrossRef]
15. Al-Warhi, T.; Sabt, A.; Elkaeed, E.B.; Eldehna, W.M. Recent advancements of coumarin-based anticancer agents: An up-to-date review. *Bioorg. Chem.* **2020**, *103*, 104163. [CrossRef]
16. Kandil, S.; Westwell, A.D.; McGuigan, C. 7-Substituted umbelliferone derivatives as androgen receptor antagonists for the potential treatment of prostate and breast cancer. *Bioorg. Med. Chem. Lett.* **2016**, *26*, 2000–2004. [CrossRef] [PubMed]
17. Meleddu, R.; Deplano, S.; Maccioni, E.; Ortuso, F.; Cottiglia, F.; Secci, D.; Onali, A.; Sanna, E.; Angeli, A.; Angius, R.; et al. Selective inhibition of carbonic anhydrase IX and XII by coumarin and psoralen derivatives. *J. Enzyme Inhib. Med. Chem.* **2021**, *36*, 685–692. [CrossRef] [PubMed]
18. Medina, F.G.; Gonzalez-Marrero, J.; Macías-Alonso, M.; González, M.C.; Córdova-Guerrero, I.; García, A.G.T.; Osegueda-Robles, S. Coumarin heterocyclic derivatives: Chemical synthesis and biological activity. *Nat. Prod. Rep.* **2015**, *32*, 1472–1507. [CrossRef]
19. Cui, J.; Li, M.-L.; Yuan, M.-S. Antifeedant activities of tutin and 7-hydroxycoumarin acylation derivatives against *Mythimna separate*. *J. Pestic. Sci.* **2012**, *37*, 95–98. [CrossRef]
20. Ji, W.; Li, L.; Eniola-Adefeso, O.; Wang, Y.; Liu, C.; Feng, C. Non-invasively visualizing cell–matrix interactions in two-photon excited supramolecular hydrogels. *J. Mater. Chem. B* **2017**, *5*, 7790–7795. [CrossRef]

21. Orhan, I.E.; Deniz, S.S.; Salmas, R.E.; Durdagi, S.; Epifano, F.; Genovese, S.; Fiorito, S. Combined molecular modeling and cholinesterase inhibition studies on some natural and semisynthetic *O*-alkylcoumarin derivatives. *Bioorg. Chem.* **2019**, *84*, 355–362. [CrossRef] [PubMed]
22. Castillo, J.-C.; Bravo, N.-F.; Tamayo, L.-V.; Mestizo, P.-D.; Hurtado, J.; Macías, M.; Portilla, J. Water-compatible synthesis of 1,2,3-triazoles under ultrasonic conditions by a Cu(I) complex-mediated click reaction. *ACS Omega* **2020**, *5*, 30148–30159. [CrossRef] [PubMed]
23. Moreno-Fuquen, R.; Arango-Daraviña, K.; Becerra, D.; Castillo, J.-C.; Kennedy, A.R.; Macías, M.A. Catalyst- and solvent-free synthesis of 2-fluoro-*N*-(3-methyl sulfanyl-1*H*-1,2,4-triazol-5-yl)benzamide through a microwave-assisted Fries rearrangement: X-ray structural and theoretical studies. *Acta Crystallogr. Sect. C Struct. Chem.* **2019**, *75*, 359–371. [CrossRef]
24. Moreno-Fuquen, R.; Hincapié-Otero, M.M.; Becerra, D.; Castillo, J.-C.; Portilla, J.; Macías, M.A. Synthesis of 1-aroyl-3-methylsulfanyl-5-amino-1,2,4-triazoles and their analysis by spectroscopy, X-ray crystallography and theoretical calculations. *J. Mol. Struct.* **2021**, *1226*, 129317. [CrossRef]
25. Shakila, G.; Periandy, S.; Ramalingam, S. Molecular structure and vibrational analysis of 1-bromo-2-chlorobenzene using ab initio HF and density functional theory (B3LYP) calculations. *J. At. Mol. Opt. Phys.* **2011**, *2011*, 512841. [CrossRef]

MDPI

Communication

New Derivatives of Lupeol and Their Biological Activity

Hoang-Thuy-Tien Le [1], Quoc-Cuong Chau [1], Thuc-Huy Duong [1,*], Quyen-Thien-Phuc Tran [1], Nguyen-Kim-Tuyen Pham [2], Thi-Hoai-Thu Nguyen [3], Ngoc-Hong Nguyen [4] and Jirapast Sichaem [5,*]

[1] Department of Organic Chemistry, Faculty of Chemistry, Ho Chi Minh City University of Education, Ho Chi Minh City 72711, Vietnam; lehoangthuytien270591@gmail.com (H.-T.-T.L.); cuongchau.dre@gmail.com (Q.-C.C.); tranphuc410@gmail.com (Q.-T.-P.T.)
[2] Faculty of Environmental Science, Sai Gon University, Ho Chi Minh City 72711, Vietnam; phngktuyen@sgu.edu.vn
[3] Faculty of Basic Sciences, University of Medicine and Pharmacy at Ho Chi Minh City, 217 Hong Bang Street, Dist. 5, Ho Chi Minh City 700000, Vietnam; nguyenthihoaithu@ump.edu.vn
[4] CirTech Institute, HUTECT University, Ho Chi Minh City 72324, Vietnam; nn.hong@hutect.edu.vn
[5] Research Unit in Natural Products Chemistry and Bioactivities, Faculty of Science and Technology, Thammasat University Lampang Campus, Lampang 52190, Thailand
* Correspondence: huydt@hcmue.edu.vn (T.-H.D.); jirapast@tu.ac.th (J.S.); Tel.: +84-919011884 (T.-H.D.); +66-54237999 (J.S.)

Abstract: The natural product lupeol (**1**) was isolated from *Bombax ceiba* leaves, which were used as starting material in the semisynthetic approach. Three new derivatives (**2a**, **2b**, and **3**) were synthesized using oxidation and aldolization. Their chemical structures were elucidated by spectroscopic analyses (HRESIMS and NMR). Compounds **3** showed significant α-glucosidase inhibition with an IC_{50} value of 202 μM, whereas **2a** and **2b** were inactive.

Keywords: lupeol derivative; benzylidene derivative; α-glucosidase inhibition; Oxone®

Citation: Le, H.-T.-T.; Chau, Q.-C.; Duong, T.-H.; Tran, Q.-T.-P.; Pham, N.-K.-T.; Nguyen, T.-H.-T.; Nguyen, N.-H.; Sichaem, J. New Derivatives of Lupeol and Their Biological Activity. *Molbank* **2021**, *2021*, M1306. https://doi.org/10.3390/M1306

Academic Editors: Giovanni Ribaudo and Rodrigo Abonia

Received: 16 November 2021
Accepted: 6 December 2021
Published: 10 December 2021

Publisher's Note: MDPI stays neutral with regard to jurisdictional claims in published maps and institutional affiliations.

1. Introduction

Diabetes mellitus (DM) causes high blood glucose after the consumption of a carbohydrate-enriched diet, leading to hyperglycemia. Uncontrolled diabetes is manifested by a very high rise in triglycerides and fatty acid levels [1]. Diverse antidiabetic drugs derived from synthetic compounds are of interest to chemists. However, these synthetic drugs come with several serious complications [1]. Due to the limitations associated with the use of existing synthetic antidiabetic drugs, the search for newer antidiabetic agents from natural sources continues. Lupeol is a pharmacologically active pentacyclic triterpenoid found in several medicinal plants worldwide [2]. It has several potential medicinal properties and is found in a variety of botanical sources [3]. Notably, lupeol has been reported to selectively target diseased and unhealthy human cells, while sparing normal and healthy cells [4]. Dozens of novel lupeol derivatives were synthesized and screened for their in vivo antihyperglycemic activity [5,6]. Most derivatives lowered the blood glucose levels, in a sucrose-challenged streptozotocin-induced diabetic rat (STZ-S) model [5]. To continue our ongoing search for highly efficient antidiabetic agents from derivatized lupeol [6,7], we herein describe the synthesis of lupeol derivatives **2**, **2a**, **2b**, and **3** (Figure 1). The structures of all the obtained compounds were characterized by ^{1}H, ^{13}C NMR, and HRESIMS. All derivatives were evaluated for α-glucosidase inhibition.

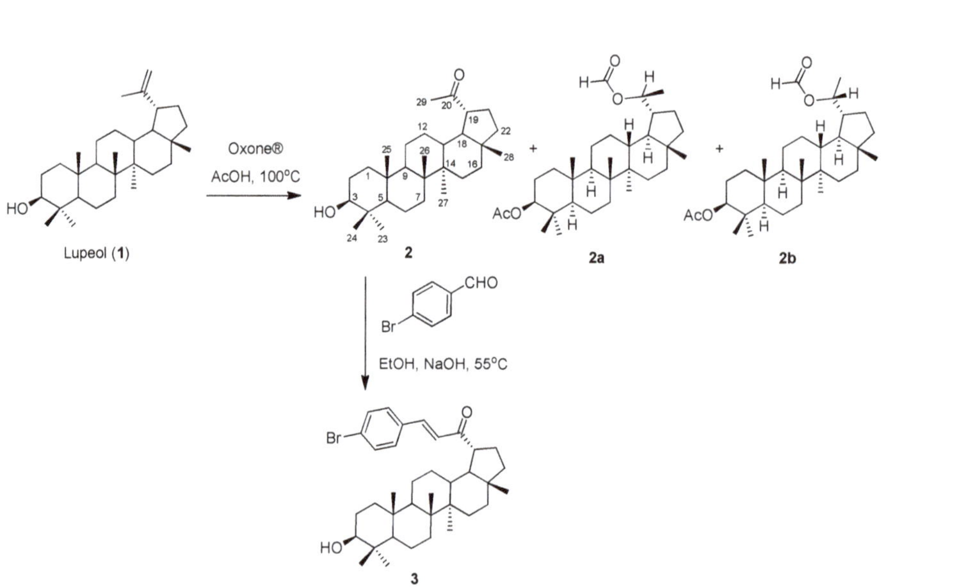

Figure 1. Synthesis of **2**, **2a, 2b**, and **3** from lupeol (**1**).

2. Results and Discussion

2.1. Synthesis

Lupeol was isolated from the Vietnamese plant *Bombax ceiba*, following our previously reported procedure [8]. Lupeol was transformed to products **2**, **2a**, and **2b** using oxidation with Oxone®, a potassium triple-salt (KHSO$_5$·1/2KHSO$_4$·1/2K$_2$SO$_4$) [6,9]. The conditions followed our previously reported method [6], with slight modifications. Both **2a** and **2b** had the same molecular formula as C$_{32}$H$_{52}$O$_4$. Comparison of NMR data of **2a/2b** and **1** indicated that oxidation occurred. The ^{1}H NMR spectrum of **2a/2b** showed differences with **1**: the downfield methine at δ_H 8.11, two oxymethines at δ_H 5.26 and 4.48, and a doublet methyl at δ_H 1.22. These signals indicated that the isopropenyl group of **1** was transformed to a 2-formylethyl group at C-19. Moreover, the downfield signal of H-3 (δ_H 4.48) indicated that 3-OH was esterified by acetic acid. The ^{13}C NMR spectrum of **2a/2b** showed one carbonyl ester at δ_C 171.1, one formyl group at δ_C 163.7 and two oxygenated carbons at δ_C 81.1 and 72.7, supporting the previous findings. Interestingly, **2a** and **2b** are C-20 epimers. Corbett and co-workers [10,11] indicated the method to define the absolute configuration of C-20 of lupane-type triterpenes. Particularly, the (20*S*) and (20*R*) isomers exhibited differences in the chemical shifts of C-19, C-20, C-29, and C-30, especially C-30. According to Corbett et al., **2a**, having C-30 at δ_C 20.1, would have a 20*R* configuration. On the other hand, **2b** would have the 20*S* configuration due to the lower chemical shift of C-30 at δ_C 14.2.

Compound **2** was further aldolized with 4-bromobenzaldehyde to afford compound **3**. Compound **3** had the same molecular formula as C$_{36}$H$_{51}$BrO$_2$, determined by a protonated ion peak at *m/z* 595.3188 in HRESIMS. Comparison of 1D NMR data of **2** and **3** indicated obvious differences. The first difference is the presence of a 1,4-disubstituted benzenoid characterized by two *ortho*-coupled protons at δ_H 7.51 and 7.42, and a *trans* double bond at δ_H 6.75 and 7.46. This was confirmed by the disappearance of a methyl ketone group at δ_H 2.15 (CH$_3$-29). This finding indicated that the aldolization occurred exclusively at C-29. The second difference was in the ^{13}C NMR spectrum. This spectrum showed the presence of seven aromatic carbons at δ_C 141.0 (C-1), 133.9 (C-5′), 132.3 (C-2′), 129.8 (C-3′,7′), and 126.9 (C-4′, 6′), supporting the reaction at C-29.

2.2. α-Glucosidase Inhibition of 2a, 2b, and 3

Compounds **2a**, **2b**, and **3** were evaluated for α-glucosidase inhibition. Only compound **3** exhibited moderate α-glucosidase inhibition with an IC_{50} value of 202 μM, compared with an acarbose-positive control (IC_{50} 360 μM). Other compounds were inactive.

3. Materials and Methods

3.1. Materials

Reagents and solvents were obtained from commercial suppliers and were used without further purification. Column chromatography was carried out using Merck Kieselgel 60 silica gel (particle size: 32–63 Å). Analytical TLC was performed using Merck precoated silica gel 60 F-254 sheets.

NMR spectroscopic data were acquired on Bruker Avance III apparatus at 500 MHz for ^{1}H NMR and 125 MHz for ^{13}C NMR. HRESIMS spectra were recorded on a Bruker MICROTOF-Q 10187.

Extraction and Isolation. The air-dried *Bombax ceiba* leaves (4 kg) were ground into powder and exhaustively extracted at room temperature with MeOH (2 × 10 L). The filtered solution was evaporated under reduced pressure to afford a residue (473.4 g). This crude extract was subsequently partitioned using solvents of *n*-hexane and EtOAc to yield *n*-hexane (40 g) and EtOAc (88 g) extracts. The *n*-hexane extract was fractionated by silica gel column chromatography (CC), eluted with *n*-hexane–EtOAc (isocratic, 10:1, *v/v*), to produce five fractions (H1-H5). Fraction H2 (15 g) was rechromatographed by silica gel CC using *n*-hexane–CHCl$_3$ (isocratic, 12:1, *v/v*) as eluent to afford lupeol (**1**) (1.5 g).

3.2. Synthesis Procedure

Synthesis of **2**, **2a**, and **2b**: Lupeol (**1**, 200 mg, 0.469 mmol) was oxidized with Oxone® (951 mg, 1.548 mmol) in acetic acid (40 mL) at 100 °C for 3 h. The mixture was stirred and continuously monitored by TLC. The mixture was extracted with EtOAc–water (1:1) to gain the organic layer. This solution was evaporated to afford a residue. Then, the residue was purified by silica gel CC to give compounds **2**, **2a**, and **2b**.

Compound **2**. Isolated yield: 74.6 mg (37%), white solid. ^{1}H and ^{13}C NMR data were consistent with those reported previously [6].

(3a*R*,5a*R*,5b*R*,7a*R*,9*S*,11a*R*,11b*R*,13a*R*,13b*S*)-1-((*S*)-1-(formyloxy)ethyl)-3a,5a,5b,8,8, 11a-hexamethylicosahydro-1*H*-cyclopenta[a]chrysen-9-yl acetate (2a). Isolated yield: 9.4 mg (4%), white solid. ^{1}H NMR (500 MHz, CDCl$_3$, δ, ppm): 2.05 (3H, s, CH$_3$-2′), 8.00 (1H, s, OCHO-29), 5.33 (1H, m, H-20), 4.48 (1H, dd, J = 10.5, 6.0 Hz, H-3), 2.13 (1H, m, H-3), 1.18 (3H, d, J = 6.5 Hz, CH$_3$-30), 1.03 (3H, s, CH$_3$-26), 0.90 (3H, s, CH$_3$-27), 0.87 (3H, s, CH$_3$-25), 0.85 (3H, s, CH$_3$-23), 0.84 (3H, s, CH$_3$-24), 0.79 (1H, d, J = 9.5 Hz, H-5), 0.76 (3H, s, CH$_3$-28). ^{13}C NMR (125 MHz, CDCl$_3$, δ, ppm): 171.2 (C-6′), 21.5 (C2′), 161.6 (C-29), 81.1 (C-3), 73.4 (C-20), 55.5 (C-5), 50.2 (C-9), 48.8 (C-18), 43.5 (C-17), 43.0 (C-14), 42.6 (C-19), 41.0 (C-8), 40.5 (C-22), 38.5 (C-4), 38.0 (C-1), 37.3 (C-13), 37.2 (C-10), 35.5 (C-16), 34.4 (C-7), 29.9 (C-21), 28.1 (C-23), 27.3 (C-15), 27.1 (C-2), 23.8 (C-12), 21.0 (C-11), 18.4 (C-6), 18.1 (C-28), 16.7 (C-25), 16.4 (C-26), 16.1 (C-24), 14.4 (C-27), 14.2 (C-30). HRESIMS calcd C$_{32}$H$_{52}$NaO$_4$ ([M+Na]$^+$): 523.3732, found: 523.3763.

(3a*R*,5a*R*,5b*R*,7a*R*,9*S*,11a*R*,11b*R*,13a*R*,13b*S*)-1-((*R*)-1-(formyloxy)ethyl)-3a,5a,5b,8,8, 11a-hexamethylicosahydro-1*H*-cyclopenta[a]chrysen-9-yl acetate (2b). Yield: 9.4 mg (5%), white solid. ^{1}H NMR (500 MHz, CDCl$_3$, δ, ppm): 2.04 (3H, s, H-2′), 8.11 (1H, s, OCHO-29), 5.26 (1H, m, H-20), 4.48 (1H, dd, J = 11.5, 5.5 Hz, H-3), 2.31 (1H, m, H-19), 1.22 (3H, d, J = 6.5 Hz, CH$_3$-30), 1.03 (3H, s, CH$_3$-26), 0.86 (3H, s, CH$_3$-25), 0.85 (3H, s, CH$_3$-27), 0.85 (3H, s, CH$_3$-23), 0.84 (3H, s, CH$_3$-24), 0.77 (1H, d, J = 2.0 Hz, H-5), 0.75 (3H, s, CH$_3$-28). ^{13}C NMR (125 MHz, CDCl$_3$, δ, ppm): 171.1 (C-6′), 21.5 (C-2′), 163.7 (C-29), 81.1 (C-3), 72.7 (C-20), 55.5 (C-5), 50.0 (C-9), 47.1 (C-18), 44.4 (C-19), 43.2 (C-17), 43.0 (C-14), 41.0 (C-8), 40.1 (C-22), 38.5 (C-4), 38.0 (C-1), 37.5 (C-13), 37.3 (C-10), 35.3 (C-16), 34.4 (C-7), 29.9 (C-21), 28.1 (C-23), 27.4 (C-15), 26.9 (C-2), 23.9 (C-12), 21.0 (C-11), 20.1 (C-30), 18.4 (C-6), 18.1 (C-28), 16.7

(C-25), 16.3 (C-26), 16.1 (C-24), 14.4 (C-27). HRESIMS calcd $C_{32}H_{52}NaO_4$ ([M+Na+H$_2$O]$^+$): 541.3870, found: 541.3869.

Synthesis of **3**: Compound **2** (70 mg, 0.163 mmol) together with NaOH (35 mg, 0.875 mmol) in ethanol (7 mL) was stirred at 55 °C for 15 min. Then, 4–bromobenzaldehyde (64.35 mg, 0.35 mmol) was added to the mixture. The reaction was performed at 55 °C for 2 h. The mixture was extracted with EtOAc–water (1:1, v/v) to gain the organic layer. This solution was applied to silica gel CC using the gradient system of *n*-hexane–EtOAc (10:1, v/v) to obtain compound **3**. Isolated yield: 68 mg (48%), white solid.

(*E*)-3-(4-bromophenyl)-1-((1*R*,3a*R*,5a*R*,5b*R*,9*S*,11a*R*)-9-hydroxy-3a,5a,5b,8,8,11a-hexamethylicosahydro-1*H*-cyclopenta[a]chrysen-1-yl)prop-2-en-1-one (**3**): ^{1}H NMR (500 MHz, CDCl$_3$, δ, ppm): 7.51 (2H, *d*, *J* = 8.5 Hz, H-3′,7′), 7.46 (1H, *d*, *J* = 16.0 Hz, H-6′), 7.42 (2H, *d*, *J* = 8.5 Hz, H-4′,6′), 6.75 (1H, *d*, *J* = 16.0 Hz, H-29), 3.19 (1H, *dd*, *J* = 11.2, 4.8 Hz, H-3), 2.87 (1H, *td*, *J* = 11.5, 6.0 Hz, H-19), 1.02 (3H, *s*, CH$_3$-26), 0.98 (3H, *s*, CH$_3$-27), 0.96 (3H, *s*, CH$_3$-23), 0.84 (3H, *s*, CH$_3$-24), 0.80 (3H, *s*, CH$_3$-25), 0.75 (3H, *s*, CH$_3$-28). ^{13}C NMR (125 MHz, CDCl$_3$, δ, ppm): 204.1 (C-20), 141.0 (C-6′), 133.9 (C-5′), 132.3 (C-3′,7′), 129.8 (C-4′,6 ′), 126.9 (C-29), 124.7 (C-2′), 79.0 (C-3), 55.4 (C-5), 50.4 (C-9), 50.1 (C-18), 43.3 (C-17), 42.9 (C-14), 40.9 (C-8), 40.3 (C-22), 39.0 (C-4), 38.8 (C-1), 37.3 (C-10), 35.2 (C-16), 34.3 (C-7), 28.7 (C-15), 28.1 (C-23), 27.9 (C-2), 27.5 (C-12), 21.1 (C-11), 18.4 (C-6), 18.3 (C-28), 16.2 (C-26), 16.0 (C-25), 15.5 (C-24), 14.6 (C-27). HRESIMS calcd $C_{36}H_{52}BrO_2$ ([M−H]$^-$): 595.3151, found: 595.3188.

3.3. α-Glucosidase Inhibitory Assay

The α-glucosidase (0.2 U/mL) and substrate (5.0 mM *p*-nitrophenyl-α-D-glucopyranoside) were dissolved in 100 mM pH 6.9 sodium phosphate buffer [12]. The inhibitor (50 µL) was preincubated with α-glucosidase; then, the substrate (40 µL) was added to the reaction mixture. The enzymatic reaction was carried out at 37 °C for 20 min and stopped by the addition of 0.2 M Na$_2$CO$_3$ (130 µL). Enzymatic activity was quantified by measuring absorbance at 405 nm. All samples were analyzed in triplicate at five different concentrations around the IC$_{50}$ values, and the mean values were retained. The inhibition percentage (%) was calculated as follows: Inhibition (%) = [1 − (A$_{sample}$/A$_{control}$)] × 100.

4. Conclusions

Three new derivatives, **2a**, **2b**, and **3**, from the natural product lupeol have been synthesized via oxidation and aldolization routes and evaluated for their α-glucosidase inhibition. Synthetic compound **3** showed much stronger α-glucosidase inhibitory activity (IC$_{50}$ 202 µM) than acarbose (IC$_{50}$ 360 µM). Synthetic products **2a** and **2b**, which lacked the 3-OH group, exhibited lower activity than **3** toward α-glucosidase. This result confirmed that this substituted group might be involved in α-glucosidase inhibition.

Supplementary Materials: The following are available online. Copies of HRESIMS and NMR spectra for compound **2a**, **2b**, and **3**.

Author Contributions: Conceptualization, T.-H.D. and J.S. methodology, T.-H.D. and J.S.; formal analysis, H.-T.-T.L., T.-H.D. and J.S.; investigation, T.-H.-T.N., H.-T.-T.L., Q.-C.C., T.-H.D. and Q.-T.-P.T.; data curation, N.-K.-T.P. and N.-H.N.; writing—original draft preparation, T.-H.D. and J.S.; writing—review and editing, T.-H.D. and J.S.; supervision, T.-H.D.; project administration, T.-H.D. All authors have read and agreed to the published version of the manuscript.

Funding: The study was funded by The Youth Incubator for Science and Technology Programme, managed by the Youth Development Science and Technology Center—Ho Chi Minh Communist Youth Union and Department of Science and Technology of Ho Chi Minh City (38/2020/HĐ-KHCNT-VU). This work was also supported by Thammasat University Research Unit in Natural Products Chemistry and Bioactivities.

Institutional Review Board Statement: Not applicable.

Informed Consent Statement: Not applicable.

Data Availability Statement: The data for the compounds presented in this study are available in the Supplementary Materials of this paper.

Conflicts of Interest: The authors declare no conflict of interest.

References

1. Tabish, S.A. Is diabetes becoming the biggest epidemic of the twenty-first century? *Int. J. Health Sci.* **2007**, *1*, 3–8.
2. Gallo, M.B.C.; Sarachine, M.J. Biological activities of lupeol. *Int. J. Biomed. Pharm. Sci.* **2009**, *3*, 46–66.
3. Starks, C.M.; Williams, R.B.; Norman, V.L.; Lawrence, J.A.; Goering, M.G.; O'Neil-Johnson, M.; Hu, J.F.; Rice, S.M.; Eldridge, G.R. Abronione, a rotenoid from the desert annual *Abronia villosa*. *Phytochem. Lett.* **2011**, *4*, 72–74. [CrossRef] [PubMed]
4. Siddique, H.R.; Saleem, M. Beneficial health effects of lupeol triterpene: A review of preclinical studies. *Life Sci.* **2011**, *88*, 285–293. [CrossRef]
5. Papi Reddy, K.; Singh, A.B.; Puri, A.; Srivastava, A.K.; Narender, T. Synthesis of novel triterpenoid (lupeol) derivatives and their in vivo antihyperglycemic and antidyslipidemic activity. *Bioorg. Med. Chem. Lett.* **2009**, *19*, 4463–4466. [CrossRef]
6. Phan, H.V.T.; Duong, T.H.; Pham, D.D.; Pham, H.A.; Nguyen, V.K.; Nguyen, T.P.; Nguyen, H.H.; Nguyen, N.H.; Sam-ang, P.; Phontree, K.; et al. Design and synthesis of new lupeol derivatives and their α-glucosidase inhibitory and cytotoxic activities. *Nat. Prod. Res.* **2020**, Article in press. [CrossRef] [PubMed]
7. Sichaem, J.; Aree, T.; Lugsanangarm, K.; Tip-Pyang, S. Identification of highly potent α-glucosidase inhibitory and antioxidant constituents from *Ziziphus rugosa* bark: Enzyme kinetic and molecular docking studies with active metabolites. *Pharm. Biol.* **2017**, *55*, 1436–1441. [CrossRef] [PubMed]
8. Sichaem, J.; Inthanon, K.; Funnimid, N.; Phontree, K.; Phan, H.V.T.; Tran, T.M.D.; Niamnont, N.; Srikittiwanna, K.; Sedlak, S.; Duong, T.H. Chemical constituents of the stem bark of *Bombax ceiba*. *Chem. Nat. Compd.* **2020**, *56*, 909–911. [CrossRef]
9. Uyanik, M.; Akakura, M.; Ishihara, K. 2-Iodoxybenzenesulfonic acid as an extremely active catalyst for the selective oxidation of alcohols to aldehydes, ketones, carboxylic acids, and enones with Oxone. *J. Am. Chem. Soc.* **2009**, *131*, 251–262. [CrossRef]
10. Corbett, R.E.; Cong, A.N.T.; Wilkins, A.L.; Thomson, R.A. Lichens and Fungi. Part 17. The synthesis and absolute configuration at C-20 of the (*R*)- and (*S*)-epimers of some 29-substituted lupane derivatives and of some 30-norlupan-20-ol derivatives and the crystal structure of (20*R*)-3β-acetoxylupan-29-ol. *J. Chem. Soc. Perkin Trans.* **1985**, *17*, 2051–2056. [CrossRef]
11. Corbett, R.E.; Cong, A.N.T.; Holland, P.T.; Wilkins, A.L. Lichens and Fungi. XVIII. Extractives from *Pseudocyphellaria rubella*. *Aust. J. Chem.* **1987**, *40*, 461–468. [CrossRef]
12. Dao, T.B.N.; Nguyen, T.M.T.; Nguyen, V.Q.; Tran, T.M.D.; Tran, N.M.A.; Nguyen, C.H.; Nguyen, T.H.T.; Nguyen, H.H.; Sichaem, J.; Tran, C.L.; et al. Flavones from *Combretum quadrangulare* growing in Vietnam and their alpha-glucosidase inhibitory activity. *Molecules* **2021**, *26*, 2531. [CrossRef] [PubMed]

MDPI

St. Alban-Anlage 66

4052 Basel

Switzerland

Tel. +41 61 683 77 34

Fax +41 61 302 89 18

www.mdpi.com

Molbank Editorial Office

E-mail: molbank@mdpi.com

www.mdpi.com/journal/molbank

.